STUDY GUIDE

JONES/CHILDERS

CONTEMPORARY COLLEGE PHYSICS

STUDY GUIDE

JONES/CHILDERS
CONTEMPORARY COLLEGE PHYSICS

JOHN SAFKO
THE UNIVERSITY OF SOUTH CAROLINA

▲▼

ADDISON-WESLEY PUBLISHING COMPANY
Reading, Massachusetts • Menlo Park, California • New York
Don Mills, Ontario • Wokingham, England • Amsterdam • Bonn
Sydney • Singapore • Tokyo • Madrid • San Juan

Reproduced by Addison-Wesley from camera-ready copy supplied by the author.

ISBN 0-201-12997-3

Reprinted with corrections June, 1990.

5 6 7 8 9 10 BA 9594939291

Preface

This study guide is designed to accompany Edwin R. Jones and Richard L. Childers *Contemporary College Physics*. The purpose of any study guide is to aid students in understanding physical concepts and problem solving techniques. This aid is provided by exercises on the concepts and worked examples of application of these concepts. Physics teachers have found from experience that practice in solving problems is the only way that most students can truly understand the basic physical concepts.

Ideally each student should have a personal tutor and should have worked examples of all possible problems and all possible ways of posing problems. Neither of these is possible, so a study guide is provided to aid students. This particular study guide was developed during a two year period while offering physics via television. Draft copies of both the text and the study guide were used by students during this period. Revisions were made to both based upon student response and success. The video lectures used were *The Mechanical Universe*© and *Beyond the Mechanical Universe*©. These tapes do not provide many examples of problem solving. For these students the study guide and the text provided the major source of worked examples.

The structure of the study guide is as follows:

1. **Practice with Important Terms**

   The text gives a list of Important Terms at the end of each chapter. An understanding of the vocabulary is necessary for a discussion of the material included in this chapter. In the long run, you will find it a great help to use the exercises in this portion of the study guide. A lack of facility with the Important Terms used in each chapter is often half the difficulty that a student has with problem solving. The practice consists of three parts:

   a). Sentences with blank space(s) for you to enter the Important Term that properly completes the sentence.

   b). The Important Terms with space for you to enter the definition of expression. There is also included the chapter number and section number within that chapter to aid you if you have to "look up" the term.

   c). The answers to part a).

2. **Comments (only in some chapters)**

   Comment sections are included in some of the chapters for two reasons:

   a). There may be comments that expand upon ideas or section of the text. These are included only if they seem to be necessary in solving some types of problems. These comments are often the material that would be covered in a classroom during normal lectures. They are included to help the non-traditional student and as reinforcement for the typical student.

   b). Comments are included to help clarify the models of the real world that are implicit in the text discussion. Scientists simplify reality by building models of the real world. These models have symbols that represent objects in the real world. The ideas expressed in the comments can be found more fully explained

in articles by D. Hestenes and others. See, for example, *American Journal of Physics*, **55**, 440 (1987) and references therein.

3. Sample Solutions

The Sample Solutions are the core of this study guide. They consist of fairly detailed solutions to problems that are similar to the problems at the end of the chapter. An attempt has be made to fully explain the symbols and laws used, draw any relevant and useful figure, and to clarify each step as fully as possible.

Most of the samples involve several steps and hopefully help the student to develop a fair understanding of the concepts needed for solving problems. The emphasis is on the quality of the examples given, rather than upon the quantity of examples. Thus these Sample Solutions are not just "putting the numbers into an equation". As a consequence of this philosophy, many of the sample solutions are for problems which are complicated. It is hoped that providing these complicated problems with their solutions will enable the student to deal with other complicated problems.

It is believed that if you will take the effort to follow each step in these samples and to ask why each was taken, that you will find problem solving less difficult. As with any study aid, you will get out of this material only what you are willing to put into it.

Several templates or alternative methods of solving the problems are presented. In general, you should always make a sketch of the problem as posed or a motion diagram. The symbols used should be clearly identified. Relevant physical laws given and substitutions shown. The final result should be given with units and the appropriate significant figures.

I wish to thank those students in the two years of television courses who have provided input, suggestions and who suffered from errors of omission as well as commission in the original material. Special thanks must also be extended to William M. Lee Jr. for his efforts to eliminate errors from the final manuscript. Any remaining errors are my responsibility. Students and instructors who have corrections or suggestions for improvements are urged to submit them to me.

John L. Safko
Columbia, SC
1989

Contents

Chapter 1
Measurement and Analysis

Important Terms

Fill in the blank with the appropriate word or words.

1. _____ Law of planetary motion says that the orbits of the planets are ellipses with the Sun at one focus.

2. The number of _____ tells us how well we know a given quantity.

3. The ratio of the arc length of a sector of a circle to the radius of that sector is a measure of the included angle of the sector in _____.

4. The statement "An imaginary line from the Sun to a moving planet will sweep out equal areas in equal times" is a statement of _____ Law.

5. The meter is an example of a base or _____ unit.

6. Kepler's _____ Law states that "the ratio of the square of a planet's period to the cube of its average distance from the Sun is a constant".

7. All physical quantities have _____.

8. The second of time is an example of a(n) _____ or fundamental unit.

9. The orbits of the planets are _____.

10. The process of putting approximate numbers into relations to get rough estimates is called _____ calculation.

11. _____ notation is designed to implicitly show the number of significant figures known for a physical quantity.

Write the definitions of the following:

12. Kepler's Third Law (1.2):

13. Significant figures (1.4):

14. Base or Fundamental unit (1.6):

15. Dimension of a physical quantity (1.6):

16. The radian (1.1):

17. ellipse (1.2):

18. order of magnitude (1.5):

19. Kepler's First Law (1.2):

20. scientific notation (1.2):

21. Kepler's Second Law (1.2):

Answers to 1–11, Important Terms

1. Kepler's First; 2. significant figures; 3. radians; 4. Kepler's Second; 5. fundamental; 6. Third; 7. dimension; 8. base; 9. ellipses; 10. order of magnitude; 11. scientific.

General Comments

This chapter is our introduction to physics and for many of you to the study of any science. Scientific knowledge can be separated into two types:

Factual knowledge — theories, models and empirical data interpreted with the models and the theories. The material in the Important Terms practice provides some examples of this factual knowledge.

Procedural knowledge — this is the collection of methods, tactics, and techniques for using and checking the factual knowledge. Procedural knowledge involves the skills needed to solve the problems that will be posed in the text and on your exams. This knowledge is sometimes referred to as the scientific method.

You must be careful not to let the need for the procedural knowledge obscure the need for the factual knowledge; nor, to believe that the factual knowledge is, by itself, physics. In this guide, we will refer to these two aspects of physics again as further examples are studied

The word **model** was used in the previous paragraphs. In physics we build models to represent the real world. These models are conceptual representations of the real world. They often are simplifications of the real world. Kepler, for example, in his orbital models ignored the other planets when studying the orbit of Mars. We now recognize that this is an oversimplification of the real world. One of the skills you will need to develop is to be able to create the correct model to use to restate problems and to pose solutions.
The models of physics are mathematical, that is, physical properties are represented by quantitative variables. As we will develop during this course, mathematical models have four components:

Names — the names assigned to the objects and the agents that interact with the object. Kepler, for example, modeled the planets and the Sun with ideal objects he called planets and Sun.

Descriptive variables — these represent the properties of the object. Kepler, for example, used the positions of the planets and Sun with respect to the then supposedly fixed stars as some of his variables. He also introduced the concept to the period of the planets. Since the Earth was one of these planets, this period had to be a period measured with respect to the fixed stars as seen from the Sun.

The **equations of the model** — these describe the structure and time evolution of the model. Kepler's Third Law provides an example of this. To Kepler the third law was an equation of his model. When we study Newtonian mechanics, we will have a model in which the laws of Kepler are deduced rather than being equations of the model.

Finally, we have an **interpretation** — the interpretation relates the descriptive variables of the model to some properties of the object which the model is representing. Often this interpretation is simple, for example, the planetary positions of Kepler are interpreted as telling us where the planet would appear in the sky. Until we make an interpretation the equations of a model represent nothing — they would be simply abstract relations among mathematical variables. An ellipse and its equation are abstract mathematical relations and concepts; however, the ellipse has meaning when we interpret it as the path through space traveled by a planet.

Each of these four aspects of a mathematical model are important. As you proceed through this course, reference will be made to these aspects.

Comments on Symbols and Equations

We will have a number of equations in our models. The equations are a mathematical shorthand way of writing sentences that relate the descriptive variables of the model. The equations, in and of themselves, have no meaning unless we know what the symbols are to represent — that is an <u>interpretation</u> must be made. Kepler's Third law is written in equation 1.2 of the text as

$$\frac{T^2}{R^3} = k.$$

This equation has meaning only when we know the interpretation of the letters T, R, and k. We could just have well written

$$\frac{P^2}{A^3} = h$$

with P being the time for a complete orbit, A the average distance of the planet from the Sun and h a constant. Both of these equations have the same physics, only the names of the variables and their interpretation has changed.

Each model usually has a set of symbols which are commonly used to represent the variables and the interpretation of those variables, such as F for force; but, the common convention need not be followed. To be safe it is best to always give the interpretation of the variables used. To repeat — unless you know what the symbols (variables) represent, you have no physics, only mathematics.

A Word On Solving Physics Problems

During this course you will have to solve problems such as you find in the text book at the end of the chapters. Simply putting down an equation, putting in numbers and getting a number is not a solution of a physics problem. It will be necessary for you to identify your model and explain the physical meaning of the variables used. You must also interpret the results in terms of the posed physical problem. The sample solutions at the end of each chapter should provide a guide.

Comments on Making Sketches

As a general procedure, you will be well advised to make a sketch of each problem you have to solve. This sketch is a visual model of the problem. The sketch does not need to be to scale or even well drawn to be useful. Of course, the better the sketch is drawn, the more helpful it may be to you.

On the sketch you can note which directions are considered positive and what symbols are used for the directions and objects involved in the problem. If motion is involved, draw a path. When you solve the model if the path length turns out be be negative, that would mean that the path is in the opposite direction from what you assumed when you made the sketch. The sketch and numbers tell you this, so you can properly interpret the results and re-draw the sketch if desired.

Conversion of Units

It will often be necessary to convert units when solving problems. This may be a conversion of units within a system of units or a conversion between two different systems. Examples of the former are millimeters to meters (SI) or inches to feet (English). An example of the latter is inches to meters (English to SI).

As a general policy, it is best to put all your initial information into the **base** or **fundamental units** of one consistent system. Usually the SI system is the best choice; although, in a few cases, problems may be posed so that the English system is preferable. This means, for example, that all distances should be expressed in meters; do not express some in meters and some in cm and others in mm..

If you adopt the policy of consistent units, you have another check on your calculations. That is, the result will come out in an expected combination of units. That is if you are calculating a quantity that is a length the answer will come out to be meters not m/s or some other unit. You should always carry the units along with your calculations and work them out separately.

Practice with Unit Conversions

Following the procedures outlined in section 1.8 of the text, work out the following conversions. The answers are given below as a check. Some helpful factors are:

$$1 = \frac{2.54 \text{ cm}}{1 \text{ in}}; \qquad 1 = \frac{3600 \text{ s}}{1 \text{ hr}} = \frac{60 \text{ s}}{1 \text{ min}} \times \frac{60 \text{ min}}{1 \text{ hr}}$$

All of these conversions are exact definitions, i.e. 1 inch is exactly equal to 2.54 cm, etc.

A. 1 yard = _____ m

B. 30 days = _____ s

C. 50 feet = _____ m

D. 3.5×10^{-3} mm = _____ m

E. $(3.0 \text{ inches})^3$ = _____ m^3

F. 1.0 ft^2 = _____ cm^2

Solutions:

A. $1 \text{ yard} \times \dfrac{3 \text{ ft}}{1 \text{ yard}} \times \dfrac{12 \text{ in}}{1 \text{ ft}} \times \dfrac{2.54 \text{ cm}}{1 \text{ in}} \times \dfrac{1 \text{ m}}{100 \text{ cm}} = 0.9144 \text{ m}$

B. $30 \text{ days} \times \dfrac{24 \text{ hr}}{1 \text{ day}} \times \dfrac{3600 \text{ s}}{1 \text{ hr}} = 2,592,000 \text{ s} \approx 2.6 \times 10^6 \text{ s}$

C. $50 \text{ ft} \times \dfrac{12 \text{ in}}{1 \text{ ft}} \times \dfrac{2.54 \text{ cm}}{1 \text{ in}} \times \dfrac{1 \text{ m}}{100 \text{ cm}} = 15.24 \text{ m}$

D. $3.5 \times 10^{-3} \text{ mm} \times \dfrac{1 \text{ m}}{1000 \text{ mm}} = 3.5 \times 10^{-6} \text{ m}$

E. $\left(3.0 \text{ in} \times \dfrac{2.54 \text{ cm}}{1 \text{ in}} \times \dfrac{1 \text{ m}}{100 \text{ cm}}\right)^3 = (7.6 \times 10^{-2} \text{ m})^3 = (7.6)^3 \times (10^{-2})^3 \times \text{m}^3$

$= (4.4 \times 10^2) \times 10^{-6} \text{ m}^3 = 4.4 \times 10^{-4} \text{ m}^3$

or

$(3.0 \text{ in})^3 = 27 \left(\text{in} \times \dfrac{2.54 \text{ cm}}{1 \text{ in}} \times \dfrac{1 \text{ m}}{100 \text{ cm}}\right)^3 = 4.4 \text{ m}^3$

F. $1.0 \left(\text{ft} \times \dfrac{12 \text{ in}}{1 \text{ ft}} \times \dfrac{2.54 \text{ cm}}{1 \text{ in}}\right)^2 = 1.0 \times (30.48 \text{ cm})^2 = 9.3 \times 10^2 \text{ cm}^2$

Note that for questions A through C the answers are exact so all the figures are significant; however, for D through F there are only 2 significant figures since the numbers 3.5, 3.0 and 1.0 appear respectively.

Sample Solutions

The method of solving these problems is the same as described in Section 1.7 of the text. We will also point out what modeling assumptions have gone into the solution.

PROBLEM SG 1.1: What is the diagonal of a rectangular appearing object whose measured sides are 37 m and 153 m ?

SOLUTION:

> **Procedure:** Consider the diagonal as forming a right triangle with two of the sides. We can have only two significant figures in the answer.

We will model the physical object with a mathematical rectangle. This mathematical rectangle has sides denoted by a and b; where for our problem $a = 37$ m and $b = 153$ m.

Figure SG 1.1 shows how a and b and the diagonal are related.

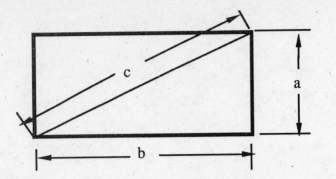

Figure SG 1.1: Rectangle of sides a and b and diagonal c.

The diagonal, c, of a rectangle is given by $c = \sqrt{a^2 + b^2}$
In terms of our model the variables are:

a — the short side of the ideal rectangle
b — the long side of the ideal rectangle
c — the diagonal of the ideal rectangle

We have solved the mathematical equation; now the numbers with units can be substituted.

$$c = \sqrt{(37 \text{ m})^2 + (153 \text{ m})^2} = \sqrt{1369 \text{ m}^2 + 23409 \text{ m}^2}$$

$$= \sqrt{24778 \text{ m}^2} = \sqrt{24778} \sqrt{\text{m}^2} = 157.4102919 \text{ m}$$

We must now interpret this model diagonal to be the diagonal of the physical object with which we started. When making the interpretation we must consider the significant figures. The quantity represented by a had two (2) significant figures and the side represented by b had three (3) significant figures. Since two (2) is the smaller number, our final answer has only two (2) significant figures. With two figures, we round 157.4... to 160, so $c = 160$ m. However, this implies that we know values to within ± 1 m; an even better form is $c = 1.6 \times 10^2$ m.

An order of magnitude calculation gives

$$c \approx \sqrt{(50 \text{ m})^2 + (150)^2} = \sqrt{(2500 + 22500) \text{ m}^2} = \sqrt{25000} \text{ m}$$

$$\approx \sqrt{4 \times 10^4} \text{ m} = 2 \times 10^2 \text{ m},$$

in agreement, to order of magnitude, with the careful calculation.

PROBLEM SG 1.2: As seen from the Earth, the Moon repeats its phases every 29.5 days (time between full Moons). This means, as seen from the Earth, the Moon appears to rotate about the Earth once each 29.5 days. How many radians/s is this revolution? Note: Contrary to most units, the radian unit is often omitted if no confusion will result. Thus if no unit is given for an angle the unit of radian is understood. Angular motions in radians/s (rad/s) are often written as "/s" or as "s⁻¹". We will explicitly put the radian in for the first few chapters.

SOLUTION:

Procedure: A complete cycle of the Moon will be 2π radians. Calculate the necessary angular velocity if the period is 29.5 days.

We replace the Moon with an idealized point body revolving about the Earth on an ideal circle. That is, we model the Moon's orbit as a circle and the motion of the Moon as uniform on this circle. We will let the symbol ω be the variable representing the radians/s for the Moon's motion. (This quantity is called the angular velocity of the Moon). We let the symbol P be the variable representing the period of the Moon's motion with respect to the Earth.

Figure SG 1.2 shows the model we are using:

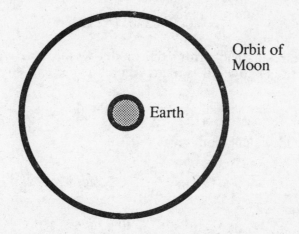

Figure SG 1.2: Orbit of Moon about the Earth.

Since there are 2π radians in a circle we expect that the angular change in radians/(time interval) times the period of rotation is the angular distance traveled — i.e. 2π radians. As an equation this statement is

$$\omega P = 2\pi,$$

so

$$\omega = \frac{2\pi}{P}.$$

We can now enter the numbers appropriate for the model, that is,

$$P = 29.5 \text{ days.}$$

$$\omega = \frac{2 \times 3.14159 \text{ radians}}{29.5 \text{ days}} = 0.21299 \text{ radians/day}$$

We must convert radians per day to radians per second. There are 60 seconds in a minute and 60 minutes in an hour and 24 hours in a day, so

$$1 \text{ day} = 24 \text{ hr} \times 60 \frac{\text{min}}{\text{hr}} \times 60 \frac{\text{s}}{\text{min}}$$

$$= (24 \times 60 \times 60) \left(\frac{\text{hr} \times \text{min} \times \text{s}}{\text{hr} \times \text{min}} \right)$$

$$= (86400)(\text{s}).$$

Thus

$$\frac{\text{day}}{86400 \text{ s}} = 1,$$

so

$$\omega = 0.21299 \frac{\text{radians}}{\text{day}} \times \frac{\text{day}}{86400 \text{ s}}$$

$$= 0.000002465162 \text{ radians/s}$$

When we pull the result back from the model to the real world there are only 3 significant figures (π is known to as many figures as needed) so the answer is

$$\omega = 2.47 \times 10^{-6} \text{ rad/s}.$$

An order of magnitude calculation is

$$\omega \approx \frac{10 \text{ rad}}{30 \text{ day}} \approx 0.3 \text{ rad/day} ,$$

and

$$1 \text{ day} \approx 20 \times 50 \times 50 \text{ s} = 5 \times 10^{4} \text{ s},$$

giving

$$\omega \approx 6 \times 10^{-6} \text{ rad/s}$$

as an order of magnitude estimate.

PROBLEM SG 1.3: A small conical shaped object has a base diameter of 5.0 inches and a height of 12.0 inches. What is its volume in cubic centimeters (cm^3) ?

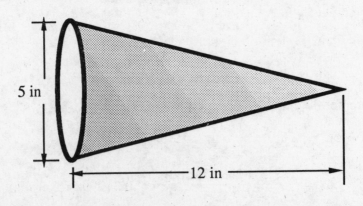

Figure SG 1.3:   Cone of base 5 inches and height 12 inches.

SOLUTION (FILL IN THE MISSING STEPS):

> **Procedure:** First convert the units to the requested centimeters and then use the formula for the volume of a cone.

The conical object is replaced by an idealized cone whose volume is given by

$$V = \frac{\pi}{3} \times r^2 \times h,$$

where r is the radius of the cone and h is its height. Label r and h on Figure 1.3:

For this problem:
$$r = \underline{\quad\quad} \text{ inches} \qquad \text{and} \qquad h = \underline{\quad\quad} \text{ inches.}$$

It is most easy to **convert units first** if more than one unit needs be converted. The conversion is
$$1 \text{ inch} = \underline{\quad\quad} \text{ cm.}$$

So

$$1 = \frac{2.54 \text{ cm}}{\text{inch}} = \frac{2.54}{1} \frac{\text{cm}}{\text{inch}} = 2.54 \frac{\text{cm}}{\text{inch}};$$

hence

$$r = 2.5 \text{ inches} \times 2.54 \times \frac{\text{cm}}{\text{inch}} = \underline{\quad\quad} \text{ cm}$$

and

$$h = \underline{\quad\quad} \text{ cm.}$$

Thus the ideal volume, V is

$$V = 1287.036997 \text{ cm}^3$$

For the physical object, the significant figures are 2 for the radius and 3 for the height, so there are only 2 significant figures in the answer. The volume is

$$V = 1.3 \times 10^3 \text{ cm}^3.$$

For an order of magnitude calculation take $V \approx r^2 h$, $r \approx 10$ cm and $h \approx 30$ cm, giving

$$V \approx 9000 \text{ cm}^3 \approx 10^4 \text{ cm}^3,$$

which is within an order of magnitude of the correct result.

PROBLEM SG 1.4: If an automobile is traveling at 55 miles/hr, what is its speed in m/s?

SOLUTION:

> **Procedure:** Convert each unit to its metric equivalent.

The conversions that are needed are:

$$1 \text{ mile} = 5{,}280 \text{ ft}; \quad 1 \text{ ft} = 12 \text{ inches}; \quad 1 \text{ inch} = 2.54 \text{ cm}$$

$$1 \text{ hr} = 60 \text{ min} = 3600 \text{ s}.$$

We will present the order of magnitude calculation first for this problem. As an order of magnitude expression

$$1 \text{ mile} \approx 6{,}000 \times 10 \times 5 \text{ cm} = 3 \times 10^5 \text{ cm} = 3 \times 10^3 \text{ m},$$

$$1 \text{ hour} \approx 4000 \text{ s}.$$

So, as an order of magnitude

$$60 \text{ miles/hr} \times \frac{3 \times 10^3 \text{ m}}{\text{mile}} \times \frac{1 \text{ hr}}{4000 \text{ s}} \approx \frac{180000 \text{ m}}{4000 \text{ s}} \approx 50 \text{ m/s}.$$

The calculation is

$$55 \frac{\text{miles}}{\text{hour}} \times \frac{5280 \text{ ft}}{\text{mile}} \times \frac{12 \text{ inch}}{\text{ft}} \times \frac{2.54 \text{ cm}}{\text{inch}} \times \frac{1 \text{ m}}{100 \text{ cm}} \times \frac{1 \text{ hr}}{3600 \text{ s}} = 24.58720 \text{ m/s}.$$

The only number with relevant significant figures is the 55 mi/hr, which has 2 significant figures. Thus

$$55 \text{ miles/hr} = 25 \text{ m/s}.$$

This number agrees with the order of magnitude calculation.

Chapter 2
Motion in One Dimension

Important Terms

Fill in the blanks with the appropriate word or words.

1. The study of motion is called _Kinematics_

2. The quantity which describes both direction and speed is _velocity_

3. The _acceleration_ of a body is the rate at which the velocity of the body is changing.

4. The horizontal axis of a graph is called the _abscissa_.

5. The study of the causes of motion is called _dynamics_

6. The _average_ speed of a body is its displacement divided by the elapsed time.

7. The _average_ acceleration is the change in velocity divided by the change in time.

8. The text value quoted for the _acceleration of gravity_, 9.8 m/s^2, is the approximate value at the earth's surface.

9. If we neglect air resistance, a body falling towards the earth is said to be in _free fall_.

10. The difference in the initial and final positions of a body is termed the body's _displacement_

11. Draw a curve of position versus time. The tangent to this curve at any time is called the _instantaneous_ velocity.

12. The vertical axis of a graph is called the _ordinate_.

13. The slope of the velocity vs displacement curve for a body is that body's _instantaneous_ acceleration.

14. The _slope_ of a line is the ratio of the change in ordinate to the change in abscissa.

15. The branch of physics which deals with kinematics and dynamics is called _mechanics_.

16. On a two-dimensional plot the axis running left to right is called the _abscissa_.

17. On a two-dimensional plot the axis running bottom to top is called the _ordinate_

Write the definitions of the following:

18. velocity (2.2):

19. distinguish between distance and displacement:

20. average acceleration (2.5):

21. kinematics (2.0):

22. dynamics (2.0):

23. slope of a line (2.3):

24. mechanics (2.0):

25. instantaneous acceleration (2.5):

26. acceleration of gravity (2.7):

27. freely falling body (2.7):

28. ordinate (2.3):

29. abscissa (2.3):

30. average velocity (2.2):

31. instantaneous velocity (2.4):

32. mechanics (2.0):

33. displacement (2.2):

34. constant speed (2.1):

35. average speed (2.1):

Answers to 1–19, Important Terms

1. kinematics; 2. velocity; 3. acceleration; 4. abscissa; 5. dynamics; 6. average; 7. average; 8. acceleration of gravity; 9. free fall; 10. displacement; 11. instantaneous; 12. ordinate; 13. instantaneous; 14. slope; 15. mechanics; 16. abscissa; 17. ordinate

Practice with Writing Equations

Let the following symbols have the interpretation given:

t	the clock time
x	position at time t
x_0	position at time t_0
v	velocity at time t
v_0	velocity at time t_0
a	acceleration — assumed constant

A bar over any symbol means:

take the average value of the quantity represented by the symbol.

The 6 most frequently used kinematic equations were given in the summary. Write them here, stopping to think carefully of the interpretation of each variable.

$$\bar{v} = \qquad\qquad\qquad \bar{a} =$$

$$x = \qquad\qquad\qquad v =$$

$$\bar{x} = \qquad\qquad\qquad v^2 =$$

In this space write what each of the symbols with a subscript mean (x_0 etc)

Comments on Using Equations

When you answered the preceding, you should have realized that the equations were written such that x_0 was the position at $t = 0$ and v_0 was the velocity at $t = 0$. We often have problems where we have position or velocity given for $t \neq 0$. It is then necessary to replace t on the right hand side of the equations by $(t - t_0)$ and to interpret x_0 as the position at t_0 and v_0 as the velocity at t_0. Equation 2.6 of the text then becomes

$$x(t) = x_0 + v_0(t - t_0) + \frac{1}{2} a (t - t_0)^2,$$

where x_0 is the position at t_0 and v_0 is the velocity at t_0.

General Comments

In this introduction to one-dimensional kinematics we meet the first and the most simple of our many models. Real bodies are idealized as point particles which we call objects (**name** for the object) which have the **variables** position, velocity and acceleration. In the real world we have a reference point from which we measure position. The usual convention is that positive positions are to the right of this point and negative positions to the left. Velocity or speed is introduced with positive velocity tending to increase the numerical value of position and negative velocities decreasing the position value (remember –5 cm is a smaller value of position than –2 cm). Likewise accelerations change the value of velocity. Positive acceleration increases velocity while negative acceleration decreases velocity.

Usually the mathematical symbols x, v, a, and t are used with the **interpretation** that x is position, v is instantaneous velocity, a instantaneous acceleration and t the elapsed time. This interpretation pulls the mathematical formulation back to the real world.

Even though we only consider constant acceleration in this chapter, we still call acceleration a variable. In this chapter acceleration is independent of time, but the acceleration can have different values depending upon the problem.

The factual knowledge is the definition of position, velocity, acceleration and the relationships that hold among them and time. The procedural knowledge is how to solve problems using these relations with prescribed accelerations.

Motion Plots and Time Plots

There are two types of drawings we can make to help understand what is happening in a given case. The first of these is shown in Fig 2.1 of the text. This is called a **motion plot.** A motion plot shows the various positions that a body has. It shows where the body has moved, not when it moved there.

The other type of plot has two axes, the horizontal axis is usually time (the independent variable) and the vertical axis can be position, velocity or acceleration. We will call such plots, **time plots**. Plots such as these allow the definition of instantaneous velocity and acceleration as a limit of the average velocity as the time intervals approach zero (See text Equation 2.3).

Both of these plots can be useful, but a motion plot should always be drawn at some point in the process of solving a problem. A comparison of the motion plot with the equations used can often help pick the approach to the solution or point out errors in your approach.

Either plot will help you keep the distinction between distance and displacement clear. The displacement is the difference between the initial and final positions without regard to the path traveled between. The distance traveled will depend upon the path traveled and may have to be calculated by summing up the displacements on various parts of the path.

Practice Reading Time Plots

Time plots are used in sections 2.4, 2.5 and 2.6 of the text to derive the idea of instantaneous velocity and acceleration. Consider the position versus time plot shown in Figure SG 2.1:

We have calculated the average velocity by the slope method for a few points. Fill in the remaining positions and/or average velocities in Table SG 2.1. Two example calculations are given beside the table.

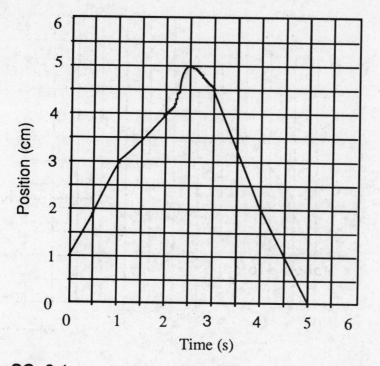

Figure SG 2.1: A position vs. time plot for a moving body.

Use the equation

$$\bar{v} = (x - x_0)/(t - t_0)$$

From the figure, we find

for $t = 1.0$ s $x = 3.0$ cm
for $t_0 = 0.5$ s $x_0 = 1.9$ cm

Table SG 2.1:
Time in seconds, position in cm, and average velocities in cm/s for Figure SG 2.1

t	t_0	x	x_0	$\bar{v}$
1.0	0.5	3.0	1.9	2.2
3.0	1.0	4.5	3.0	0.75
5.0	4.0			−2.0
2.0	1.0			
4.0	3.0			
3.0	2.0			

$\bar{v} = (3.0 - 1.9)$cm$/(1.0 - 0.5)$s
$= (1.1/0.5) \times$ (cm/s)
$= 2.2$ cm/s

for $t = 3.0$ s, $x = 4.5$ cm
for $t_0 = 1.0$ s, $x_0 = 3.0$ s

$$\bar{v} = \frac{(4.5 - 3.0)\text{cm}}{(3.0 - 1.0)\text{s}}$$

$$= 0.75 \text{ cm/s}$$

All numbers have been rounded to 2 significant figures.

The instantaneous speed can be calculated from a time plot by the method of tangent lines. Figure SG 2.2 shows the same position versus time curve as in Fig SG 2.1. The tangent lines at $t = 1$ s and at $t = 3$ s are shown and the slope of these lines are calculated. Examine this process and then repeat for the indicated values.

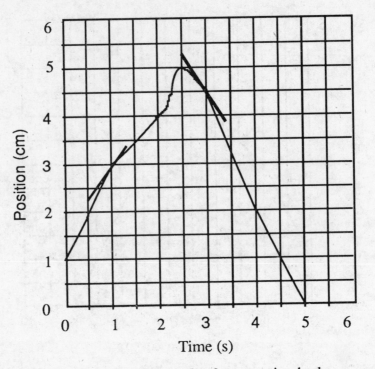

Figure SG 2.2: A position vs. time plot for a moving body.

To draw the tangent line, draw a line which just touches the position vs time curve as shown in the examples. A length of the tangent line is drawn for your convenience. Pick two points on the line and get the x and t values. then calculate the slope by the rule:

$$\text{slope} = \Delta x/\Delta t.$$

This slope will be the instantaneous velocity.

For $t = 1$ s the tangent line drawn includes the points:

$$(x,t); \ (3.0,1.0) \ \text{and} \ (2.25,0.5).$$

The slope is then

$$v(1 \ s) = \text{slope at 1 s} = \frac{\Delta x}{\Delta t} = \frac{(3.0 - 2.25)\text{cm}}{(1.0 - 0.5)\text{s}}$$

$$= \frac{0.75}{0.5} \times \frac{\text{cm}}{\text{s}} = 1.5 \ \text{cm/s}.$$

we could have used

$$v(1 \ s) = \frac{(2.25 - 3.0)\text{cm}}{(0.5 - 1.0)\text{s}} = 1.5 \ \text{cm/s}$$

with no change in the result.

Either the first x and t can be subtracted from the second x and t or the second x and t can be subtracted from the first x and t, but do the same for both x and t to get Δx and Δt

— i.e. if you have the pair (x_1, t_1) and (x_2, t_2) then

$$\Delta x = (x_1 - x_2) \text{ and } \Delta t = (t_1 - t_2)$$

OR

$$\Delta x = (x_2 - x_1) \text{ and } \Delta t = (t_2 - t_1)$$

BUT NOT

$$\Delta x = (x_1 - x_2) \text{ and } \Delta t = (t_2 - t_1)$$
[THIS CHOICE GIVES THE WRONG SIGN]

This velocity (the instantaneous velocity) is not the same as the average velocity we calculated before using the same t values. This is to be expected whenever the curve representing the motion is not a straight line.

For $t = 3$ s the tangent line drawn includes the points (3.0,4.5) and (2.5,5.5). The slope then gives

$$v(3 \text{ s}) = \frac{\Delta x}{\Delta t} = \frac{(4.5 - 5.5)\text{cm}}{(3.0 - 2.5)\text{s}} = -2.0 \text{ cm/s}.$$

The minus sign (−) tells us that the value of the position is decreasing (moving to smaller values), which is what we see on the plot.

Exercises:

Show that the velocity at 2.5 s is zero and find the velocity at 2.0 s and at 4.0 s. You may get slightly different values, since graphical analysis involves the judgement as to exactly where to put the tangent line.

A usual difficulty for beginning students is confusing displacement, velocity and acceleration when examining time plots. There are several steps that must be taken when you are confronted with a time plot.

1. Examine the graph to see what is plotted against what. Is the plot a position versus time (a displacement plot) or a velocity versus time or an acceleration versus time plot.
2. What scales are used for the plot axes. Look at the numbers and the legend — if position, is it in cm, m, feet, km? If velocity, is it in m/s, radians/s, etc? Is the lower left corner of the plot zero or some other number?
3. What can you say about the motion from the plot? If it is a position versus time plot, does the velocity change magnitude or sign? If it is a velocity versus time plot, does the acceleration change magnitude or sign?

In Figure SG 2.3a is shown the position versus time plots for two different objects, one we call red the other blue. First examine Figure SG 2.3a only:

The red object starts at 0 cm and at the end of 10 s is at 50 cm. The blue object starts at 20 cm and at the end of 10 s is at 40 cm. Consider the following questions (write your answers and check them before reading further):

Exercise:

a) At time equals 9 s, which has the larger velocity red or blue?
b) At time equals 3 s, which has the larger velocity red or blue?
c) Do the red and blue objects ever have the same speed in the interval 0–10 s?

There are several common and easy to make errors when you first start to read graphs:

    An easy error to make when you start to read graphs is to confuse the height of a line with its slope.

    A common error is to say the blue object's speed is larger at 3 s since the blue curve is higher at that time.

    Another common error is to say the speeds are the same at 7 s since the curves cross there.

In Figure SG 2.3a, the height of the line gives us the displacement while the slope gives the velocity. The slopes of the two curves are given in Figure SG 2.3b. Clearly the red object always has the larger velocity.

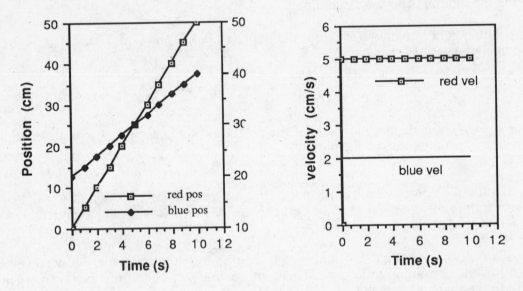

    Figure SG 2.3:   Motion of two objects. a) Position versus time plot.
                                                             b) Velocity versus time plot.

Another skill you need in time plot reading is to be able to construct the motion that is associated with a given time plot ;and conversely, given a moving body or a description of the motion of a body, you should be able to construct a time plot description of that motion.

Figure SG 2.4a shows the position as a function of time for a moving body. When the curve is rising, the body's velocity is positive; when the curve is falling, the body's velocity is negative. Thus at point A, we see that the slope of the position curve is zero, so the velocity is zero. Before the time A, the velocity is positive; just after A, it is negative as we can see in Figure SG 2.4b. The velocity changes from negative to positive between points C and D.

Looking at Figure SG 2.4a, what is the velocity at B ? At E ? Use Figure SG 2.4b for help if needed.

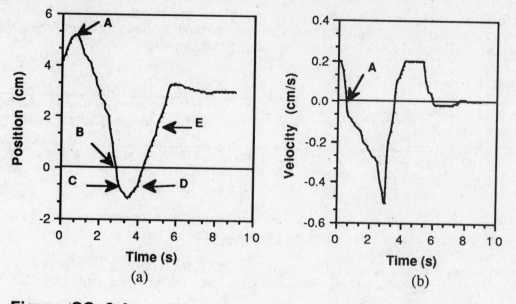

Figure SG 2.4:   The motion of a body.   a) Position versus time.
b) Velocity versus time.

Finally, a motion plot for Figure SG 2.4a would look like Figure SG 2.5.

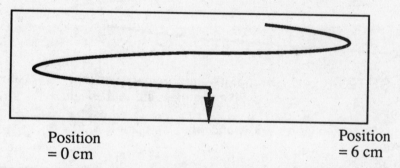

Position
= 0 cm

Position
= 6 cm

Figure SG 2.5:   Rough Motion Plot for the Time Plot Figure SG 2.4a. The
vertical displacement is just so we can see the true displacement given by the horizontal axis.

Answers to time plot questions

Table SG 2.1: 3rd line 0.0, 2.0; 4th line 4.0, 3.0, 1.0; 5th line 2.0, 4.5, – 2.5.
Figure SG 2.2: (your answers may differ slightly) v(2 s) = 1.6 cm/s; v(4 s) = – 2.4 cm/s.
Figure SG 2.3: red; red; never.
Figure SG 2.4: negative; positive.

Sample Solutions

PROBLEM SG 2.1: What is the average speed and the average velocity for a trip which is made by slowly traveling for two hours at a constant speed of 10 km/hr and then for 3 hours at a constant speed of 30 km/hr in the opposite direction?

SOLUTION:

> **Procedure:** Describe the model and draw a motion plot. The average speed is the total distance traveled divided by the elapsed time and the average velocity is the net displacement divided by the elapsed time.

We first model this problem by letting t represent the time since the trip started, x the distance traveled and v the velocity. We will assume the initial motion is to the +x direction (This is an arbitrary choice which will affect the sign of the answer).

An important aspect of our model is that we will ignore the acceleration that must occur when the velocity changes from 10 km/hr to the right to 30 km/hr to the left

A **motion plot** will make the discussion simpler. The motion plot first drawn is a preliminary version which does not really tell if we end up to the right or left of the starting point.

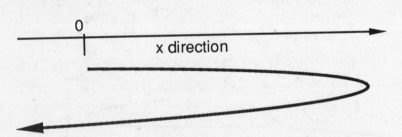

Figure SG 2.6   Motion plot — preliminary version. $x = 0$ is the starting point and positive distances are to the right.

We need to decide what we know and what we want to find.

GIVEN:

time (t)	velocity (v)
0 to 2 hrs	10 km/hr
2 hrs to 2 hrs + 3 hrs = 5 hrs	−30 km/hr

The minus sign for $t = 2$ hrs to $t = 5$ hrs is needed since the motion is now in the negative x-direction.

FIND:

the average speed $\bar{s} = \dfrac{\text{distance traveled}}{t_f - t_i}$

the average velocity $\bar{v} = \dfrac{x_f - x_i}{t_f - t_i}$,

where t_f denotes $t = 5$ hr and t_i denotes $t = 0$ hr.

We see that in order to get either average we will need to find the final displacement at t = 5 hr. We can do this if we calculate the displacement while traveling 10 km/hr to the right and the displacement while traveling 30 km/hr to the left. The final displacement is the sum of these two, keeping in mind the directions.

We will use the relation

$$x = x_0 + v_0(t - t_0) + \frac{1}{2} a (t - t_0)^2$$

where $a = 0$.

For

$$0 \leq t \leq 2 \text{ hr}, x_0 = 0 \text{ and } v_0 = 10 \text{ km/hr}$$

so

$$x(2 \text{ hr}) = 0 + (10 \text{ km/hr}) \times (2 \text{ hr}) = 20 \text{ km}$$

Then for

$$3 \text{ hr} \leq t \leq 5 \text{ hr}, t_0 = 2 \text{ hr}, x_0 = 20 \text{ km and } v_0 = -30 \text{ km/hr}$$

Thus

$$x(5 \text{ hr}) = 20 \text{ km} + (-30 \text{ km/hr}) \times (5 \text{ hr} - 2 \text{ hr})$$

$$= 20 \text{ km} + (-30 \times 3) \times \frac{\text{km} \times \text{hr}}{\text{hr}}$$

$$= 20 \text{ km} - 90 \text{ km} = -70 \text{ km}$$

We can now draw our final motion diagram as shown in Figure SG 2.7:

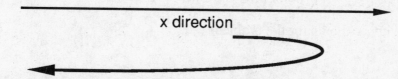

x direction

Figure SG 2.7 Motion diagram for example

The average speed is given by

$$\bar{s} = \frac{\text{distance traveled}}{t_f - t_i} = \frac{20 \text{ km} + 90 \text{ km}}{5 \text{ hrs}} = 22 \frac{\text{km}}{\text{hr}},$$

while the average velocity is given by

$$\bar{v} = \frac{x_f - x_i}{t_f - t_i} = \frac{-70 \text{ km} - 0 \text{ km}}{5 \text{ hr} - 0 \text{ hr}} = -14 \frac{\text{km}}{\text{hr}}$$

The average speed is always positive, while the negative sign of the average velocity depends upon our choice of the initial direction being called the +x direction. You should repeat this calculation with the initial velocity to the left (towards − x) and show that you obtain 14 km/hr for the average velocity.

NOTE: This problem clearly shows the difference between distance traveled and displacement. It also shows that average velocity may be misleading since the true velocity was never the average value except at the turn around.

PROBLEM SG 2.2: A ball is thrown straight upward at an initial speed of 10 m/s. How long does it take for the ball to return to its starting position? At what time and how far away from the starting point is its velocity zero and the acceleration non-zero?

SOLUTION:

> **Procedure:** This is a constant acceleration problem, so use the relation for constant acceleration. Make a motion plot.

We model the ball with a point and assume the motion is one dimensional. The problem is stated such as to assume the motion occurs in the Earth's gravitational field. Choose upward as a positive direction, z. Take $z = 0$ to be the initial starting point. Figure SG 2.8 shows the motion diagram.

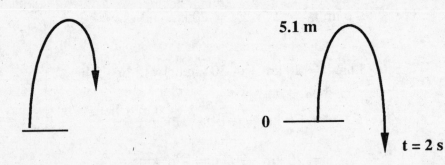

Figure SG 2.8: Initial motion diagram for Problem SG 2.2

Figure SG 2.9: Final motion diagram for Problem SG 2.2

Examine the items known for the problem

GIVEN

initial position	z_0	=	0.0 m
initial velocity	v_0	=	10.0 m/s
acceleration	a	=	-9.8 m/s^2

Since we want a time,

$$z(t) = z_0 + v_0 t + \frac{1}{2} a t^2 \qquad \text{(Equation 2.6 of text)}$$

Since we can take $t = 0$ at the start of the motion, we may use the unmodified form of the text equation.

This gives

$$z(t) = 10 \text{ m/s} \times t - 4.9 \text{ m/s}^2 \times t^2.$$

For $z(t) = 0$ we have

$$z(t) = 10 \text{ m/s} \times t - 4.9 \text{ m/s}^2 \times t^2$$

$$= 0 = t(\ 10 \text{ m/s} - 4.9 \text{ m/s}^2 \times t).$$

This has solutions

$$t = 0 \text{ s}$$

and

$$t = \frac{10 \text{ m/s}}{4.9 \text{ m/s}^2} = 2.041 \text{ s}.$$

The first time ($t = 0$ s) is the starting time and the second time ($t = 2.0$ s) would be the return time. Note that we have now rounded to 2 significant figures.

To answer the second part of the question we note that the acceleration is constant so the acceleration is never zero. However from the motion diagram, we see that the velocity is zero at the top of the motion since the direction changes. We can get the z value for this from

$$v^2 = v_0^2 + 2a(x - x_0). \text{ (Equation. 2.7 of the text).}$$

For $v = 0$

$$v^2 = 0 = (10 \text{ m/s})^2 + 2 \times (-9.8 \text{ m/s}^2) \times (\ z - 0\ ),$$

this gives

$$z = 5.1 \text{ m}.$$

So the ball would rise with decreasing velocity to a height of 5.1 m, stop and then fall down. It would pass its starting point 2 s after it started.

To calculate the time at which the ball is at rest (but accelerated). Using Equation 2.6 again, we find that t must satisfy

$$5.1 \text{ m} = (10 \text{ m/s}) \times t - (4.9 \text{ m/s}^2) \times t^2,$$

which can be rewritten as:

$$t^2 - 2.04 \text{ s} \times t - 1.04 \text{ s}^2 = 0.$$

This can be solved using the binomial theorem (see page SG 36)

The equation

$$a\, x^2 + b\, x + c = 0,$$

has the solution $x = \dfrac{-b \pm \sqrt{b^2 - 4\, a\, c}}{2a}$,

so

$$t = (1.02 \pm 0.00) \text{ s} = 1.02 \text{ s} = 1.0 \text{ s}.$$

The ball will be at rest 1 s after leaving the ground at a height of 5.1 m.

We can also use symmetry to calculate this last time. Since the entire motion takes 2.0 s, we expect the turn around to occur half way through the motion.

PROBLEM SG 2.3: A person starts from rest and moves with a constant acceleration of -2 m/s^2 for 11 seconds and then moves with an acceleration of 2 m/s for another 11 seconds: a). What is the maximum speed attained? b). What is the average speed? c). How far does the person go during the whole trip?

SOLUTION:

> **Procedure:** Make a motion plot. Use the velocity and position relations for constant acceleration,.

Model the motion with the variable x which measures the displacement from the starting position. x is positive to the right and the acceleration is assumed to change instantaneously from -2 m/s to $+2$ m/s. A preliminary motion diagram is shown in Figure SG 2.10:

-242 m

-121 m

Figure SG 2.10: Preliminary motion diagram for Problem SG 2.3

Figure SG 2.11: Final motion diagram for Problem SG 2.3

What do we know?

GIVEN:

Initial acceleration	a_i	-2 m/s, lasting for 11 seconds
Final constant acceleration	a_f	$+2$ m/s, lasting for 11 seconds.

WANT TO KNOW:

maximum speed	v_{max}
average speed	$v_{average}$
final displacement	x_{final}

The person starts from rest at $t = 0$ s, decreases her velocity (that is, the velocity has a larger negative value) until $t = 11$ s, then increases her velocity for the next 11 s. Using the relation

$$v = v_0 + a t,$$

we have

$v = -2 \ (m/s^2) \times t$	for	$0 < t < 11$ s
$v = -22$ m/s	for	$t = 11$ s
$v = -22$ m/s $+ 2(m/s^2) \times (t - 11s)$	for	11 s $< t < 22$
$v = 0$ m/s	for	$t \geq 22$ s

From this we see that the answer to (a),

the maximum speed (speed is the magnitude of velocity) is 22 m/s.

To answer (b) we need to find the final displacement. Use equation 2.6 of text for this

$$x = x_0 + v_0 t + \frac{1}{2} a t^2.$$

Since the acceleration changes at 11 seconds, we will have to break the motion into two parts.

$$x = \frac{1}{2} \times (-2 \text{ m/s}^2) \times t^2 \qquad \text{for} \qquad 0 < t < 11 \text{ s}$$

$$x = -121 \text{ m} \qquad \text{for} \qquad t = 11 \text{ s}$$

$$x = -121 \text{ m} - 22 \text{ m/s} \times (t - 11 \text{ s})$$
$$+ \frac{1}{2} \times (2 \text{ m/s}) \times (t - 11 \text{ s})^2 \qquad \text{for} \qquad 11 < t < 22 \text{ s}$$

where the -22 m/s at 11 s was found in part (a)

$$x = -121 \text{ m} - 242 \text{ m} + 121 \text{ m}$$
$$= -242 \text{ m} \qquad \text{for} \qquad t \geq 22 \text{ s}$$

Note that even though she ends up at rest, there is a net displacement.

The symmetry and equal times of the accelerations will not bring us back to the initial starting position. Assuming that she would return to the initial starting position is a common and easily made mistake.

We can now draw our final motion diagram as in Figure SG 2.8, and calculate the average velocity using Equation 2.2:

$$v_{\text{average}} = \frac{-242 \text{ m} - 0 \text{ m}}{22 \text{ s} - 0 \text{ s}} = -11 \text{ m/s}.$$

Part (c) can also be answered now. Since the velocity is always negative, the total distance traveled is 242 m and equal to the displacement. Had she changed directions, the total distance traveled would not be the displacement. We would have to add (neglecting sign) the individual parts of her trip to get the total distance traveled.

PROBLEM SG 2.4: A stone is thrown downward at 5.0 m/s from a height of 30 meters. At the same time, from directly below, a stone is thrown upward from the ground with a speed of 20 m/s. At what height do their paths intersect?

SOLUTION:

Procedure: Draw a motion diagram. To avoid confusion use a different variable for each stone, but measure both variables from the same starting position and in the same direction.

We model the two stones as points and measure all distances positively upwards from the ground. We will use the symbol t to represent time measured from zero when both stones start out on their paths. The position of the stone starting downward will be denoted by $x(t)$ and the position of the stone starting upward by $y(t)$. This is shown in Figure SG 2.12. In the figure we have assumed that the stones meet before the upward moving stone starts to fall back. The calculations are independent of this assumption.

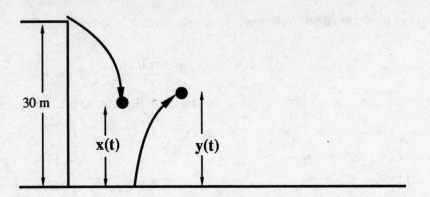

Figure SG 2.12: The two stones and their motion. The downward thrown stone's position is given by $x(t)$ and the upward thrown stone's position by $y(t)$. The motions are shown as curved paths. We assume the paths cross.

Thus $x(t)$ is the height above the ground of the downward moving stone and $y(t)$ is the height above the ground of the upward moving stone. We are asked to find the height when their paths intersect. That is we want to find the x or the y for which $x(t) = y(t)$. Since this is a constant acceleration problem, we expect Equation 2.6 of text to be the needed relation. What are we given?

GIVEN:

Stone 1

position at $t = 0$ is 30 m $x_0 = 30$ m
velocity at $t = 0$ is -5.0 m/s (minus since up is $+$) $v_{0x} = -5$ m/s
acceleration is -9.8 m/s^2

Stone 2

position at $t = 0$ is 0 m $y_0 = 0$ m
velocity at $t = 0$ is $+20$ m/s $v_{0y} = +20$ m/s
acceleration is -9.8 m/s^2

So

$$x(t) = 30 \text{ m} - (5 \text{ m/s}) \times t - (4.9 \text{ m/s}^2) \times t^2$$

and

$$y(t) = (20 \text{ m/s}) \times t - (4.9 \text{ m/s}^2) \times t^2$$

If we set these two equal [$x(t) = y(t)$], we can solve for t, the time that the paths cross, and then use either equation to find the height.

Setting $x(t) = y(t)$ gives:

$$30 \text{ m} - (5 \text{ m/s}) \times t - (4.9 \text{ m/s}^2) \times t^2 = (20 \text{ m/s}) \times t - (4.9 \text{ m/s}^2) \times t^2$$

or

$$30 \text{ m} - (5 \text{ m/s}) \times t = (20 \text{ m/s}) \times t.$$

This gives

$$t = \frac{30 \text{ m}}{25 \text{ m/s}} = 1.2 \text{ s.}$$

So
$$x(1.2 \text{ s}) = y(1.2 \text{ s}) = 24 \text{ m} - 7.1 \text{ m} = 17 \text{ m}.$$

Hence the two stones would pass 1.2 s after we started and would be at a height of 17 m.

Note that it is often necessary to find information, in this case t, that is not requested in order to be able to find the desired quantities.

Has the initially upward moving body reached its maximum height? To find this use the expression for velocity, $v(t) = v_0 + a\,t$, to find when $v(t) = 0$. For the initially upward moving body:

$$v(t) = 0 = 20 \text{ m/s} - 9.8 \text{ m/s} \times t,$$

whose solution is

$$t = 2.0 \text{ s}.$$

Figure SG 2.12 was drawn correctly.

PROBLEM SG 2.5: A boat is traveling on a river whose current flows at 5.0 miles/hr. The boat is capable of traveling 12.0 miles/hr in still water. The boat travels downstream for 1.2 hr. The boat then turns quickly around and returns to its starting position. What is the total duration of the trip?

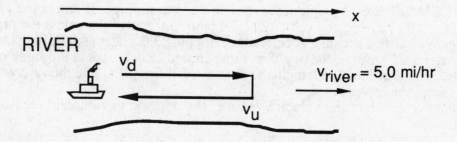

Figure SG 2.13: Boat moving at constant speed with respect to a flowing river.

SOLUTION:

Procedure: Since this problem is posed in the English system and an answer with the units of hours will be acceptable, we will not convert the initial data to the SI system. Since the velocity of the boat is given relative to the water, this is a problem in velocity addition.

We first make a sketch of the problem as shown in Figure SG 2.13. The velocity of the boat with respect to the water is 12 mi/hr so if we let v_d be the downstream velocity and v_u be the upstream velocity:

and
$$v_d = 12.0 \text{ mi/hr} + 5.0 \text{ mi/hr} = 17 \text{ mi/hr}$$

$$v_u = -12.0 \text{ mi/hr} + 5.0 \text{ mi/hr} = -7.0 \text{ mi/hr}$$

where we have chosen the downstream direction as positive.

We take the initial position of the boat to be $x = 0$ at $t = 0$.

So for

$$0 < t < 1.2 \text{ hr}, \qquad x(t) = v_d \, t$$

and for

$$1.2 \text{ hr} < t < t_f \qquad x(t) = x_0 + v_u \, (t - t_0)$$

where

$$x_0 = v_d \times 1.2 \text{ hr} = 20 \text{ mi}$$

and

$$t_0 = 1.2 \text{ hr.}$$

During the upstream portion of the trip

$$x(t) = 20 \text{ mi} - 7.0 \text{ mi/hr} \times (t - 1.2 \text{ hr}).$$

The boat returns to the starting position when t = t_f and x(t_f) = 0. This gives

$$x(t) = 0 = 20 \text{ mi} - 7.0 \text{ mi/hr} \times (t_f - 1.2 \text{ hr}),$$

or

$$t_f = 1.2 \text{ hr} + \frac{20 \text{ mi}}{7.0 \text{ mi/hr}} = 4.1 \text{ hr.}$$

The trip takes 4.1 hr and the boat travels 20 mi downstream before returning to its starting position.

PROBLEM SG 2.6: A bug crawls into an open drain pipe. It enters the pipe at a time we call t = 0.0 s moving at a velocity of 2.0 cm/s. 5.0 s after she enters the pipe, she starts slowing down with a constant acceleration, stopping 3.0 s later (8.0 s after starting into the pipe). After a 1.0 s pause, the bug then turns around, accelerates at the same rate as before towards the end of the pipe which she first entered for 4.0 s. She then moves at constant velocity until she leaves the pipe.

(a). Make a sketch of the velocity vs time, the position vs time and the acceleration vs time.
(b). What is the acceleration and what is the final velocity of the bug and at what time does she leave the pipe.

SOLUTION:

> **Procedure:** The first part of this problem is a practice in translating a word description into a motion plot. Since we are given the time periods of acceleration, we will have to start with an acceleration plot, add that up to get a velocity plot and then construct the position plot.

To analyze this problem it is helpful to make a table of known quantities and of the unknowns. The columns will be the time intervals, and the rows the position, velocity and acceleration.

Table SG 2.2

time (s)	position (cm)	velocity(cm/s)	acceleration (cm/s^2)
0.0	0.0	2.0	0
0.0 < t < 5.0	$x_1(t)$	2.0	0
5.0 < t < 8.0	$x_2(t)$	$v_2(t)$	a
8.0 < t < 9.0	x_3	0.0	0
9.0 < t < 13.0	$x_4(t)$	$v_4(t)$	a
13.0 < t < t_f	$x_5(t)$	v_5	0

Examine Table SG 2.2. It contains all the information you were given about the bug's motion. For the first 5.0 s the velocity is a constant 2.0 cm/s. For the next three seconds the acceleration is a constant which we call a. The motion stops by $t = 8.0$ s and the position remains fixed for 1.0 s. Then the same acceleration acts for 4.0 s and the velocity is constant until the bug leaves the pipe. This information alone is sufficient to allow us to sketch the acceleration, the velocity, and the position as functions of time.

Let us take the initial direction of motion as positive. Then the acceleration is a negative value between 5.0 and 8,0 seconds and between 9.0 and 13.0 seconds and is zero for all other times. This is shown in Figure SG 2.14(a).

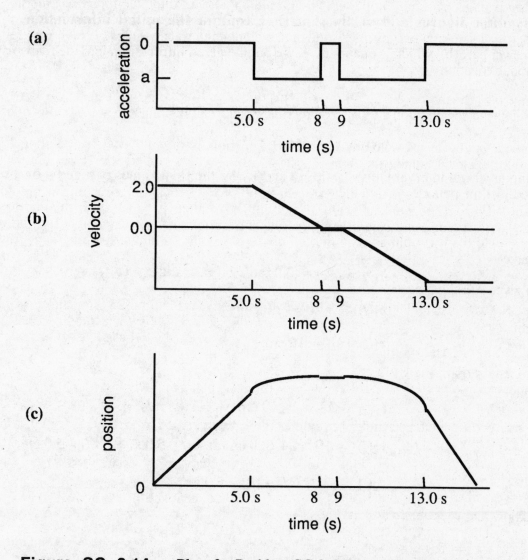

Figure SG 2.14: Plots for Problem SG 2.6: (a) acceleration vs time. (b) velocity vs time. (c) position vs time.

The velocity starts at +2.0 cm/s, remains constant until $t = 5.0$ s; then, since the acceleration is constant, it decreases linearly until $t = 8.0$ s when the velocity is zero. The velocity remains zero until $t = 9.0$ s. It then decreases linearly (with the same slope as for 5 s $< t <$ 8 s) until $t = 13.0$ s. After that time the velocity is constant. The velocity is negative after $t = 9.0$ s and the slopes for the two intervals of acceleration must be the same since the accelerations are equal. This is shown in Figure SG 2.14(b).

The position must start at 0.0 for $t = 0.0$ s, move upward with a constant slope (slope = velocity) until $t = 5$ s. The position curve then bends over to a fixed value of position for 8.0 s $< t <$ 9.0 s.
The position curve must then bend down until $t = 13.0$ s after which the slope is constant. The slope of the final straight line portion is negative while the initial slope is positive. Since the acceleration time on the return path is longer than during the first part of the motion, the final slope does not have the same magnitude as the initial slope. This is shown in Figure SG 2.14(c).

Study these figures and verify that they contain the initial information

To answer part (b), which is quantitative, we put the given information into the relevant kinematic equation:

$$x(t) = x_0 + v_0(t - t_0) + \frac{1}{2} a(t - t_0)^2$$

and

$$v(t) = v_0 + a(t - t_0).$$

During any time interval the values of x_0 and v_0 are the final values of x and v of the preceding time period.

For
$$0.0 \text{ s} < t < 5.0 \text{ s},$$
we have
$$t_0 = 0, v_0 = 2.0 \text{ cm/s}, x_0 = 0, \text{ and } a = 0.$$
This gives
$$x_1(t) = v_0 t = 2.0 \text{ cm/s} \times t,$$
hence
$$x_1(5.0 \text{ s}) = 10 \text{ cm}.$$
For
$$5.0 \text{ s} < t < 8.0 \text{ s},$$
we have
$$t_0 = 5.0 \text{ s}, x_0 = 10 \text{ cm}, v_0 = 2.0 \text{ cm/s},$$
so
$$x_2(t) = 10 \text{ cm} + (2.0 \text{ cm/s}) \times (t - 5.0 \text{ s}) + \frac{1}{2} a (t - 5.0 \text{ s})^2$$
and
$$v_2(t) = 2.0 \text{ cm/s} + a(t - 5.0 \text{ s}).$$

We know that v_2 is zero at

$$t = 8.0 \text{ s},$$
so
$$v_2(8 \text{ s}) = 2.0 \text{ cm/s} + a \times 3.0 \text{ s} = 0.$$
Hence
$$a = -0.67 \text{ cm/s}^2,$$
and
$$x_3 = x_2(8 \text{ s}) = 10 \text{ cm} + 6.0 \text{ cm} - 1.0 \text{ cm} = 15 \text{ cm}.$$

Thus for
$$9.0 \text{ s} < t < 13.0 \text{ s},$$
we have
$$x_0 = 15 \text{ cm}, v_0 = 0 \text{ and } a = -0.67 \text{ cm/s}^2.$$
So
$$v_4(t) = a(t - 9.0 \text{ s}) = -0.67 \text{ cm/s}^2 (t - 9.0 \text{ s})$$
and
$$x_4(t) = 15 \text{ cm} - 0.33 \text{ cm/s}^2 (t - 9.0 \text{ s})^2.$$
When
$$t = 13.0 \text{ s},$$
these give
$$v_4(13.0 \text{ s}) = -2.7 \text{ cm/s and } x_4(13.0 \text{ s}) = 4.7 \text{ cm}.$$
Hence,
$$v_5 = v_4(13.0 \text{ s}) = -2.7 \text{ cm/s},$$
and
$$x_5(t) = 4.7 \text{ cm} - 2.7 \text{ cm/s} (t - 13.0 \text{ s}).$$

The bug exits the pipe when $x_5(t) = 0$. This occurs when

$$t = 13.0 \text{ s} + \frac{4.7 \text{ cm}}{2.7 \text{ cm/s}} = 14.7 \text{ s} = 15 \text{ s}.$$

So the acceleration is -0.67 cm/s^2, the final velocity is -2.7 cm/s and the bug leaves the tube 15 s after entering.

PROBLEM SG 2.7: An automobile starts from rest and reaches a speed of 100 km/hr in 10.0 seconds. Find the average acceleration. What distance would be covered if the acceleration is constant?

SOLUTION:

Procedure: Use the definition of average acceleration. Then use the displacement law to find the distance traveled.

If we let a_{avg} be the average acceleration, Δv the change in velocity and Δt the elapsed time,

$$v_{avg} = \frac{\Delta v}{\Delta t}$$

$$= \frac{(100 - 0) \text{ km/hr}}{10 \text{ s}}$$

$$= \frac{100 \text{ km/hr} \times 10^3 \text{ m/km} \times 3600 \text{ s/hr}}{10 \text{ s}} = 2.78 \text{ m/s}^2$$

If this average acceleration is the true constant acceleration, we may write the position $x(t)$ as

$$x(t) = x_0 + v_0 t + \frac{1}{2} a t^2,$$

where x_0 is the initial position and v_0 is the initial velocity (both assumed zero). Using the average acceleration we find at when $t = 10$ s, displacement $= 139$ m.

The last step may also be calculated from

$$v^2 = v_0^2 + 2ax .$$

Solving this for x gives, since $v_0 = 0$,

$$x = \frac{v^2}{a} = 139 \text{ m}.$$

The average acceleration is 2.78 m/s^2 and the displacement after 10 s is 139 m.

Chapter 3
Motion in Two Dimensions

Important Terms

Fill in the blanks with the appropriate word or words.

1. A quantity having magnitude only is called a(n) _____.

2. The sum of two vectors is called the _____ vector.

3. The distance a projectile travels from its launch on the earth until it hits the earth is called the _____.

4. A(n) _____ has both magnitude and direction.

5. The _____ velocity of two moving bodies is just the vector difference of the individual velocities.

Write the definitions of the following:

6. scalar (3.0):

7. vector (3.0):

8. relative velocity (3.2):

9. resultant (3.1):

10. range of a projectile (3.6):

Answers to 1–5, Important Terms

1. scalar; 2. resultant; 3. range of a projectile; 4. vector; 5. relative.

General Comments

This chapter introduces both factual knowledge and procedural knowledge.
The factual knowledge is the idea of vectors, the parabolic motion of projectiles and the relation for the range of a projectile. The procedural knowledge is how to combine vectors (both graphically and by components), the independent treatment of the components of motion, and the posing and solution of projectile problems.

This chapter builds on the last chapter. Since we can treat individual components separately the equations from Chapter 2 apply to each component.

Addition of Vectors

Vectors can be added two ways:

        I. graphically
       II. adding components.

We will consider both ways in the following examples. These examples are also used to show you how we model reality with figures. Since our examples involve trips of several km, we could not possibly draw the figure to full scale, so we need a scaling factor. The first example will show how the scaling is used to convert the figure to the real world that we are modeling with the graph.

EXAMPLE 1

A student makes a trip which has two parts as shown in Figure SG 3.1. First he traveled to the North-Northeast as shown and then to the East as shown. The scale is 1/2 inch = 2 km. The parts of the trip are labeled A and B.

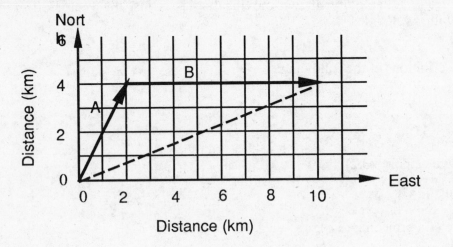

Figure SG 3.1:   Two part trip.

The parts of the trip can be defined either as a magnitude and a direction or as two components. With a ruler and a protractor, measure the lengths and directions. Measure lengths to the nearest 1/10 inch (or nearest 1/16 and round the fraction to the 1/10) and measure angles to the nearest 0.5 degree. For example the length of vector A measures 2 and 3/16 inch; 3/16 = 0.18 so A is 2.2 inches.
Since 1/2 inch = 2 km, we can write

$$1 = \frac{2 \text{ km}}{1/2 \text{ inch}} = 4 \text{ km/inch}.$$

Multiply the measured distance in inches by this conversion factor (4 km/inch) to get the represented length in km.

Read the scales to get components. Also draw a vector **C** which is the sum of **A** and **B** and get its magnitude, direction and components. Check your answers with the entries and verify that the components of **C** are the algebraic (signed) sums of the components of **A** and **B**.

	Magnitude			Component Northern		Eastern	
Vector	inches	km	Direction	inch	km	inch	km
A	1.1	4.4	26°.5	1.0	4.0	0.5	2.0
B	2.0	8.0	90°	0.0	0.0	2.0	8.0
C	2.7	10.8	75°	1.0	4.0	2.5	10.0

EXAMPLE 2

Repeat the above procedure for the 3 vector sum shown in Figure SG 3.2. Also draw the vector $D = A + B + C$. Calculate the scale factor and measure length and directions. All directions are measured from North again. Check your entries against the answers and reconcile any major differences. The scale is

$$\frac{1}{2} \text{ inch} = 1 \text{ km}.$$

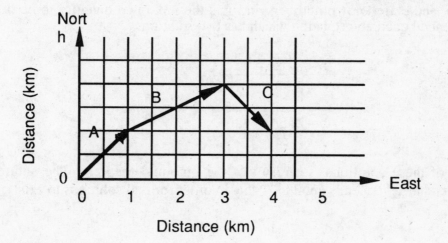

Figure SG 3.2:   A vector sum of 3 vectors.

Scale factor : 1 = (___ km/ ____ inch) = _____ km/inch

	Magnitude			Component Northern		Eastern	
Vector	inches	km	Direction	inch	km	inch	km
A				0.5	1.0		
B							
C							
D							

Answers to EXAMPLE 2: (scale factor 1 inch = 2 km)

A: 0.70 in = 1.4 km, 45°, 0.5 in=1 km N, 0.5 in = 1 km E;
B: 1.1 in = 2.2 km, 60°.5, 0.5 in = 1 km N, 1.0 in = 2 km E;
C: 0.70 in = 1.4 km, 135°, –0.5 in = –1 km N (– means S), 0.50 in = 1 km E;
D: 2.1 in = 4.2 km, 76°, 0.5 in = 1 km N, 2 in = 4 km E.

These examples provided practice with calculating vector sums using the two different methods (graphical analysis and components). Which method is best to use will depend upon the particular problem under consideration.

Solving Quadratic Equations

Beginning in this chapter and in many of the following chapters you will need to solve equations of the form:

$$a\, x^2 + b\, x + c = 0$$

where a, b, and c are known numbers and x is the unknown quantity to be determined. This is a general quadratic equation which has two solutions

$$x_1 = \frac{-b + \sqrt{b^2 - 4ac}}{2a}$$

and

$$x_2 = \frac{-b - \sqrt{b^2 - 4ac}}{2a}.$$

The form of these solutions is a result of the binomial theorem. The nature of these solutions, x_1 and x_2, depends upon $b^2 - 4ac$. In order for real solutions to exist

$$b^2 - 4ac \geq 0.$$

If this is not true there are no real numbers that satisfy the equation. For a physics problem this means one of two things:

1. The physical behavior is different than assumed.

2. You have made a mistake, either mathematically or in the physical quantities you used.

In general, for a physics problem, you will find two real solutions (or only one real solution if $b^2 - 4ac = 0$). It is possible that only one of the solutions is physically acceptable. There is no general principle to tell you which is the acceptable solution. Each problem must be analyzed to determine which solution is the one you want by asking if the solution makes physical sense. For example if x must physically or because of initial conditions be greater than zero, a negative x should be discarded. Another possibility is that one of the solutions implies that two solid bodies must pass though each other. By asking such questions, you should be able to determine which solutions to keep and which to discard.

Sample Solutions

PROBLEM SG 3.1: A boat capable of traveling 20 km/hr in still water is traveling downstream in a river flowing at 5.0 km/hr. A student who can float but not swim falls overboard (or maybe is thrown) and is not missed for 30 minutes. The boat turns around and heads upstream. How long from the turn around time until the student is found?

SOLUTION:

> **Procedure:** Use the relation between velocity and time. Find the time when the boat and the student have the same position. We will solve the problem with two different choices of references.

CHOICE 1; Measure from where the student fell overboard:

Since the question deals with a boat on a moving river, it is suggested that the problem is one of relative velocity. It is probably best to describe everything with respect to the river bank so we can keep sums straight. The simplest units seem to be km and hours. but other units are also possible. Since all the motion is along the river we will let x represent the position of the boat measured from the point the student fell (was pushed) overboard, y the position of the student from water entry point and t will be the elapsed time measured from this event. We will choose positive x and y in the direction of river flow. Let v_B be the velocity of the boat relative to the water and v_R the velocity of the river relative to the bank.

GIVEN:

velocity of boat relative to water $\qquad \left(\begin{array}{l} \text{20 km/hr downstream} \\ \text{–20 km/hr upstream} \end{array} \right)$

velocity of river relative to bank $\qquad$ 5.0 km/hr

velocity of student relative to water $\qquad$ 0.0 km/hr

> Some comments on the givens: Notice we will have to change the sign of the velocity of the boat depending on its motion. + if downstream, – if upstream. We also need to recognize that a floating, non-swimming student moves with the water.

A preliminary motion diagram is shown in Figure SG 3.3.

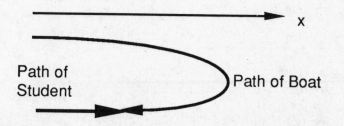

Figure SG 3.3   Motion Diagram for Student and Boat in Problem SG 3.1.

The path of the boat is in two parts, downstream and upstream, while the student only moves downstream. It is reasonable to neglect the turn around interval for no other reason than that we have no information. During the downstream portion of the trip the velocity of the boat relative to the river bank is 20 km/hr + 5 km/hr = 25 km/hr, while the upstream velocity is –20 km/hr + 5 km/hr = –15 km/hr. Equation 2.6 relates position and time and there is no acceleration.

$$x(t) = x_0 + v_0 t .$$

We will need to find how far the boat goes before it turns around to get the value of x_0 to use in the preceding equation for the second part of the trip. Since t is measured from when $x = 0$, we will also need to subtract the 0.5 hr from the t on the right hand side of the equation.

For the boat

$x(t) = (25 \text{ km/hr}) \times t$	for	$0 \leq t \leq 0.50$ hr
$x(0.5 \text{ hr}) = 12.5$ km	for	$t = 0.50$ hr
$x(t) = 12.5 \text{ km} – (15 \text{ km/hr}) \times (t –.50 \text{ hr})$	for	$0.50 \text{ hr} \leq t \leq t_{end}$

The first step finds the value of x_0 used in the equation for the last part of the trip. For the floating student.

$$y(t) = (5 \text{ km/hr}) \times t \qquad \text{for} \qquad 0 \leq t \leq t_{end}$$

The boat and student meet at t_f when $x(t_f) = y(t_f)$. Setting $x(t) = y(t)$ gives

$$x(t_f) = 12.5 \text{ km} – (15 \text{ km/h}) \times (t_f – 0.50 \text{ h}) = (5 \text{ km/h}) \times t_f = y(t_f)$$

$$t_f = \frac{12.5 \text{ km} + 7.5 \text{ km}}{20 \text{ km/hr}} = 1 \text{ hr}$$

The problem asks how long after turn around is it before the student is picked up, so the answer is $t_f – 0.50 \text{ hr} = 30$ min.

CHOICE 2; Measure from the turn around point for the boat:
We could have also solved the problem choosing the origin of x and y to be the turn around point and measuring both x and y in the upstream direction. We will also measure t from the turn around. In this case the choice will make the problem a bit longer; but the problem can still be solved. We will solve the problem this way in order to show you that you don't have to make a lucky guess of origins or coordinate directions to solve a problem.

GIVEN:

river velocity	–5 km/hr
boat velocity for $t < 0$	–25 km/hr
boat velocity for $t > 0$	15 km/hr
student velocity for all t	– 5.0 km/hr

Then, for the boat

$$x(t) = (–25 \text{ km/hr}) \times t \qquad \text{for} \qquad t < 0 \text{ hr}$$

$$x(t) = (15 \text{ km/hr}) \times t \qquad \text{for} \qquad 0 \text{ hr} < t < t_{end} ,$$

while for the student

$$y(t) = y_0 + (-5.0 \text{ km/hr}) \times (t - t_0) \qquad \text{for} \qquad t_0 < t < t_{\text{end}}$$

We know that the student fell off when $t = -0.50$ hr, so $t_0 = -0.50$ hr. But what is y_0? We can get y_0 by realizing that there are two times when the student and boat are together — when she fell off and when we pick her up. This means that

$$x(t) = y(t) \qquad \text{when} \qquad t = -0.50 \text{ hr,}$$

so

$$x(-0.50) = (-25 \text{ km/hr}) \times (-0.50 \text{ h}) = 12.5 \text{ km} = y(-0.50 \text{ hr}) = y_0 .$$

Thus

$$
\begin{aligned}
y(t) &= 12.5 \text{ km} + (-5.0 \text{ km/hr}) \times (t + 0.50 \text{ hr}) \\
&= 10 \text{ km} - 5 \times t \times (\text{km/hr})
\end{aligned}
$$

Now we can find the time that the boat and the student are together again since for t_f

$$x(t_f) = y(t_f) .$$

This gives

$$t_f = 0.50 \text{ hr} = 30 \text{ min.}$$

PROBLEM SG 3.2: A stone is thrown over a cliff with a velocity of 20 m/s and at an upwards angle of 45°. If the cliff is 25 m high, how far from the base of the cliff will the stone hit?

SOLUTION:

> **Procedure:** Draw a motion plot. The motion in the vertical and the horizontal direction is independent. First find when the stone hits the ground; then determine the horizontal distance traveled during this time interval.

The usual first step in any two or more dimensional problem is to draw the motion picture as in Figure SG 3.4.

We will idealize the cliff to a vertical wall and assume the stone was thrown from the exact top and edge. We will let $x(t)$ represent the distance in the horizontal direction from the cliff's edge as shown, with $t = 0$ being the throwing of the stone. We will also let $y(t)$ be the height of the stone above the ground measured positively upwards. Any air resistance will be neglected.

The approach is to use the independence of the two components of the position and velocity to write two separate equations. for each quantity

The stone hits the ground when $y(t) = 0$. If we can find the time for which this is true, we can then put that value of t into $x(t)$ to find the distance traveled in the x–direction.

First we need to express the initial velocity in x and y components.

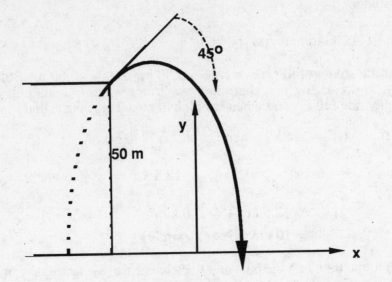

Figure SG 3.4:   The thrown stone.

From the drawing we see that the initial x velocity is 20 m/s × cos(45°) = 14 m/s in the positive x-direction. The initial y-velocity is 20 m/s × sin(45°) = 14 m/s upwards (positive y).

GIVEN

quantity	symbol	value
initial x position	x_0	0.0 m
initial y position	y_0	50.0 m
initial x velocity	v_{xo}	+14.0 m/s
initial y velocity	v_{yo}	+14.0 m/s
x acceleration	a_{xo}	0.0 m/s^2
y acceleration	a_{yo}	−9.8 m/s^2

The relevant model equations are then

$$x(t) = v_{xo}\, t = (14 \text{ m/s}) \times t$$

$$y(t) = y_0 + v_{yo}\, t + \frac{1}{2}\, a\, t^2$$

$$= 50 \text{ m} + (14 \text{ m/s}) \times t - (4.9 \text{ m/s}^2) \times t^2.$$

Now, using the binomial theorem for $y(t) = 0$
for

$$t = \frac{-14 \text{ m/s} \pm \sqrt{(14 \text{ m/s})^2 - 4 \times (-4.9 \text{ m/s}^2) \times (50 \text{ m})}}{-9.8 \text{ m/s}^2}$$

$$= 4.9 \text{ s} \quad \text{or} \quad -2.1 \text{ s}.$$

Since the stone hits the ground for a time $t > 0$,

$$t = 4.9 \text{ s.}$$

Then using this value of t

$$x(4.9 \text{ s}) = (14 \text{ m/s}) \times 4.9 \text{ s} = 69 \text{ m.}$$

So the stone hits the ground 4.9 s after launch and at a distance of 69 m from the cliff. Only 2 significant figures have been kept in agreement with the initial data.

The negative value of t (-0.23 s) given by the binomial theorem is the time at which the stone would have been launched from a level at the base to pass the true launch point with the given velocity (assuming the cliff were not in the way). This path is shown as a dotted line on Figure SG 3.4.

As an aside we can use

$$v(t) = v_0 + a \, t$$

to find the maximum height of the stone.. When the stone is at the top of its flight the y velocity is zero, so

$$0 = 14 \text{ m/s} + (-9.8 \text{ m/s}^2) \times t.$$

That is, the maximum occurs at a time $t = 1.4$ s. At $t = 1.4$ s,

$$y(1.4 \text{ s}) = 50 \text{ m} + 19.6 \text{ m} - 9.6 \text{ m} = 60 \text{ m}$$

and

$$x(1.4 \text{ s}) = 20 \text{ m}$$

PROBLEM SG 3.3: A space explorer fires his gravity testing projectile at an angle of 35° with respect to the horizontal and a muzzle velocity of 50 m/s and finds that the range is 500 m. What is the acceleration of gravity of the planet?

SOLUTION:

Procedure: Since the launch and impact point are at the same height, we can use the range formula Equation 3.8.

First draw a motion diagram as in Figure SG 3.5. Also draw the initial velocity vector at 35°.

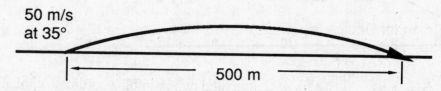

50 m/s
at 35°

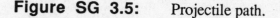

500 m

Figure SG 3.5: Projectile path.

We model the projectile as a point object and measure range along the horizontal. We know that the range of a projectile, R, is related to the launch angle, θ, the initial or muzzle velocity, v_0, and the local gravity, g, by Equation 3.8 of text

$$R = v_0{}^2 \sin(2\theta)/g.$$

For this problem

$$v_0 = 50 \text{ m/s}, \theta = 35°,$$

and g is to be found.

So

$$g = v_0{}^2 \sin(2\theta)/R = (50 \text{ m/s})^2 \times \sin(70°)/500 \text{ m} = 4.698 \text{ m/s}^2.$$

So the acceleration of gravity on this planet is 4.7 m/s^2 or about half that of the earth.

PROBLEM SG 3.4: A stone is thrown horizontally off a 50 m high building. It hits the ground at a distance of 20 m from the base of the building. What is the initial horizontal velocity or the stone? Neglect air resistance. How much longer would it take the stone to hit the ground if it were initially thrown with twice the horizontal speed?

SOLUTION:

> **Procedure:** Use the property that the vertical and horizontal motions are independent. To show the independence of the choice of origins, we will solve this problem with two choices.

Let x be the horizontal coordinate and y the vertical coordinate. We will use x_0 for the initial x position, y_0 for the initial y position, v_{xo} for the initial x-velocity, $v_{yo} = 0$ for the initial y-velocity and g to represent the acceleration of gravity.

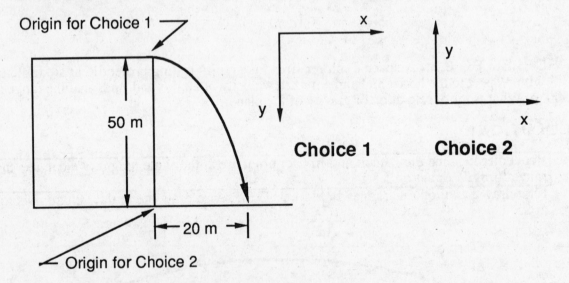

Figure SG 3.5: Sketch of motion and two possible coordinate choices.

FIRST COORDINATE CHOICE

We will choose a coordinate system with an origin at the top of the building and with x measured positively to the right and y positively downwards. This is shown in the box labeled CHOICE 1 of Figure SG 3.6. For this choice

$$x_0 = 0 \qquad y_0 = 0 \qquad v_{xo} = ? \qquad v_{yo} = 0 \qquad g > 0.$$

Then the general equations of motion are

$$y(t) = y_0 + v_{yo} t + \frac{1}{2} a t^2 = \frac{1}{2} g t^2$$

and

$$x(t) = x_0 + v_{xo} t + \frac{1}{2} a t^2 = v_{xo} t.$$

with g a positive number (particles are accelerated in the +y direction).

We are given that when y = 50 m, x = 20 m. This occurs at a time t_f such that

$$y(t_f) = 50 \text{ m} = \frac{1}{2} g t_f$$

and

$$x(t_f) = 20 \text{ m} = v_{xo} t_f.$$

This gives

$$t_f = \sqrt{\frac{100 \text{ m}}{g}} = 3.2 \text{ s}$$

and

$$v_{xo} = \frac{20 \text{ m}}{t_f} = 6.3 \text{ m/s}.$$

This coordinate choice is unusual in as much as it is a left handed coordinate system. That is if you rotate +x into +y by the smallest angle, the rotation is clockwise rather than counterclockwise. This does not matter for 2-dimensional problems, but when we deal with 3-dimensional problems you may get into trouble with the equivalent 3-dimensional left handed system. The definitions of certain quantities (angular momentum and magnetic field) depend upon the right hand nature of the coordinate system.

SECOND COORDINATE CHOICE

Choose the coordinate origin at the foot of the building with x increasing positively to the right and y increasing in an upward direction as shown in Choice 2 of Figure SG 3.6. For this choice

$$x_0 = 0 \qquad y_0 = 50 \text{ m} \qquad v_{xo} = ? \qquad v_{yo} = 0 \qquad g < 0.$$

Then

and

$$x(t) = v_{xo} t$$

$$y(t) = y_0 + \frac{1}{2} g t^2.$$

The stone hits the ground at a time t_f given by

$$y(t_f) = 0 = y_0 + \frac{1}{2} g t_f^2.$$

This gives

$$t_f = \sqrt{\frac{100 \text{ m}}{-g}} = 3.2 \text{ s},$$

and

$$v_{xo} = \frac{20 \text{ m}}{t_f} = 6.3 \text{ m/s}.$$

This is the same result as before as expected.

Since the x and y motion are independent, the time it takes the stone to hit the ground is independent of the initial horizontal velocity.

Chapter 4
Force and Motion

Important Terms

Fill in the blanks with the appropriate word or words.

1. The gravitational force is believed to depend upon the _____ of the distance.

2. Newton's Laws of motion are valid in a(n) _____ reference frame.

3. A force of one _____ acting on a 1 km mass will give that mass a 1 m/s^2 acceleration.

4. When an object is sitting on a table at rest, the force produced by the table is perpendicular and just the amount needed to balance gravity. This force is called the _____ force.

5. The _____ is the constant of proportionality between the magnitude of the kinetic frictional force and the magnitude of the normal force.

6. Newton's _____ Law relates the change in momentum of a body and the applied force.

7. The property of matter that resists changes in motion is called _____.

8. The _____ is a measure of the amount of matter in a body.

9. "In the absence of outside forces a body has constant velocity " is a statement of Newton's _____ Law.

10. "If a body exerts a force **F** on another body then the second body exerts a force –**F** on the first" is a statement of Newton's _____ Law.

11. _____ equilibrium occurs when a body is at rest and the sum of all forces acting on it are zero.

12. The force produced by two surfaces sliding across each other is called _____.

13. _____ equilibrium occurs when a body has constant velocity and the sum of all forces acting on it are zero.

14. _____ occurs when a body freely falls in a gravitational field.

15. The vector diagram for a body showing all forces acting on the body is called a(n) _____.

16. The _____ of a body is the gravitational force exerted upon that body.

17. The product of the mass and the velocity of a body is called the _____ of the body.

18. The vector sum of all the forces acting on a body is called the ____ force on the body.

19. The _____ of gravity on the earth's surface is about 9.0 m/s^2.

20. If a mass is hung on a wire, the force in the wire that opposes the force applied by the mass is termed the _____ in the wire.

Write the definitions to the following:

21. difference between mass and weight (4.5):

22. net force (4.2):

23. Newton's Third Law (4.6):

24. momentum (4.4):

25. newton (4.4):

26. static equilibrium (4.9):

27. dynamic equilibrium (4.9):

28. normal force (4.5):

29. weight (4.5):

30. mass (4.4):

31. free body diagram (4.7):

32. friction (4.8):

33. acceleration of gravity (4.5):

34. inertial reference frame (4.10):

35. inverse square force (4.1):

36. inertia (4.3):

37. coefficient of friction (4.8):

38. Newton's First Law (4.3):

39. Newton's Second Law (4.4):

40. weightlessness (4.6):

41. tension (4.7):

Answers to 1–20, Important Terms

1. inverse square; 2. inertial; 3. newton; 4. normal; 5. coefficient of friction; 6. Second; 7. inertia; 8. mass; 9. First; 10. Third; 11. static; 12. friction; 13. dynamic; 14. weightlessness; 15. free body diagram; 16. weight; 17. momentum: 18. net: 19. acceleration; 20. tension

General Comments

This chapter introduces a number of new **variables** into our model of the real world. The mass is an intrinsic property of each body. When we have problems where the mass changes, we can break the body into smaller pieces, each of which has its own intrinsic mass, or we can relax the intrinsic nature of mass and let it vary just as the velocity and position do. Variables which change are position, velocity and acceleration which we had in Chapter 3. Finally, there are the forces which are variables representing interactions among the bodies of the problem.

We have finally introduced some **equations of the model**, Newton's Laws. These laws provide relationships among the variables for both single bodies (Newton's First Law) and among the bodies themselves (Newton's Second and Third Laws). Although the three laws of Newton as given in this text and most others use the word **body**, we must remember that so far we have a model where all bodies are point objects. We will later see how to extend or modify our model to include extended bodies.

We have also introduced several new objects into our model. One such object is the surface. The surface is an idealized plane or curved object which has the property of being impenetrable and perfectly rigid. The surface of our model is clearly an object which does not exist in the real world; however, we will interpret the surface as our modeling of a real but simplified surface.

Our model surface can also have a property that we call the coefficient of friction. This is our modeling of the real frictional properties of surfaces. The coefficient of friction depends both upon the surface and upon the object resting or moving upon the surface. For any problem we have a given body resting on a given surface, so there is a number (usually denoted by the Greek letter μ), which gives the coefficient of friction. We will discuss friction more fully under its own separate heading.

Other objects introduced are the light inextensible string and the massless pulley. The string whose mass can be neglected and whose length does not change is a model of the real world string. Such a string is assumed for now to transmit tensional forces along its length without loss and not to support any compressional forces. That is, if you pull on the string with a force F, this force appears without loss at the opposite end.

Likewise the ideal pulley is the model for the real world pulley which is used to change the direction of forces without introducing the complication of friction. Both of these idealizations can be considered a good approximation when the masses of the attached bodies are large compared with the string and pulley masses and for the pulley when the friction of the bearings can be neglected. In later chapters both the string and pulley will be given additional properties.

Comments on Friction

In this model, we take the frictional force as a given expression without attempting to understand how it arises in the real world. As indicated in the text, the frictional force can be understood in terms of molecular forces and Newton's Laws. In the context of the point object model used at this stage in the course, we will have to take the friction as simply a given property. The frictional force is a force which always opposes motion along the surface upon which the body is moving or resting. In terms of Figure SG 4.1 we take the direction x to be along the surface .

If you use the free body diagram approach, you must consider the tension in the string as a force applied to each body and which is horizontal in SG 4.1a and at an angle θ in SG 4.1b. Even though we are modeling the body as a point, figures usually show it as extended for visibility.

In Fig SG 4.1a the surface is perpendicular to the weight of the body, so the normal force, N, is opposite and equal to the weight, mg (note the weight of an object is a vector). This equality follows from Newton's Second Law and the unstated assumption that the body does not move off the surface. As a vector expression

$$N = - mg.$$

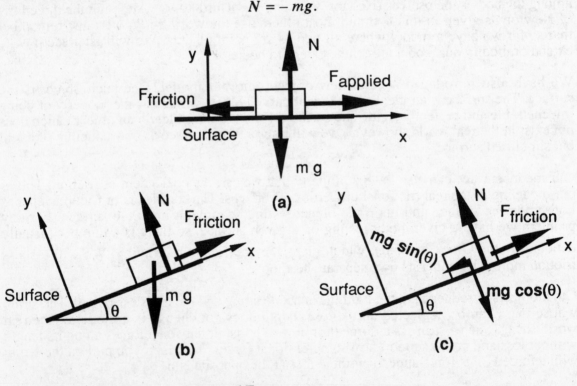

Figure SG 4.1:   Frictional Force:
a) horizontal surface with an external applied force $F_{applied}$;
b) tilted surface;
c) Force components resolved along x and y.

If no other forces act ($F_{applied} = 0$ in Fig SG 4.1a) then the frictional force is zero. If there is an applied force, there will be a frictional force opposing the applied force. If the component of the applied force parallel to the surface is large enough, the body will begin to accelerate and the frictional force opposing the motion will be given by μN. Note that any component of the applied force perpendicular to the surface will change the value of the normal force and hence affect the size of the frictional force. See Problem SG 4.4 for an example.

In Figure SG 4.1b we show the same body on an inclined plane and the forces that act. In Figure SG 4.1c the gravitational force is decomposed into its x and y-components. The normal force, N, is in the $+y$ direction and has magnitude mg cos θ. The body is in equilibrium provided the frictional force, $F_{friction}$, can equal the x-component of gravity. This is possible provided

$$F_{friction} = - (-mg \sin \theta) = mg \sin \theta \le \mu N = \mu\, mg \cos \theta$$

which means that

$$\mu \ge \tan \theta.$$

If the preceding inequality is not satisfied, then the block is accelerated down the incline by a force $-mg \sin \theta + \mu N < 0$ (the $-$ sign indicates acceleration down the incline).

In no case does the frictional force encourage acceleration. It can only oppose applied forces and reduce the resulting acceleration. For the case shown in Figure SG 4.1b–c, the applied force along the plane is $-mg \sin \theta$, provided by the force of gravity. Only when the applied force is larger than the maximum frictional force (μN), does the body move.

Comments on Force Diagrams

The drawing of one or more force diagrams is essential to the understanding of, and the solution of, problems. Force diagrams should supplement motion diagrams, not replace them. Of course, if the problem is a statics problem, (sum of forces equals zero), a motion diagram will be uninformative. In addition to the force diagrams, it is often useful to draw a separate figure showing the physical arrangement.

There are two kinds of force diagrams introduced in physics texts — the external force diagram and the free body diagram. The use of each of these is shown at the start of text Section 4.7 with the Atwood machine example. We can not replace the real Atwood machine with a single point body. The best we can do is to model each mass as a point and replace the string with an idealized massless inextensible string. If we ask only about the motion of the system of two masses and string, then, using the external force diagram, it is possible to solve for the motion of the system and not find the tension in the string connecting the masses.

If you use the free body diagram approach, you must consider the tension in the string as a force applied to each body and also have the tension at each end of the string be equal and opposite. The free body diagram approach has the advantage of forcing you to consider all forces that are acting, even forces that hold separate parts (non-point) of the system together. However, this method has the disadvantage of forcing you to find forces which need not be known to solve the posed problem.

Although, the free body diagram approach may take a few extra steps; studies have shown that a consistent free body diagram approach leads to better student understanding of the posed problems. We will use the free body approach for all sample solutions as well as the external force approach when appropriate.

Sample Solutions

PROBLEM SG 4.1: Four coplanar forces act on a 10 kg mass. Three of them are as follows (directions are measured clockwise relative to the reference direction):

      7.00 N at 0°;   20.0 N at 100°;   25.0 N at 250°·

What is the magnitude and direction of the fourth force if the 10 kg mass is to be in translational equilibrium?

SOLUTION:

> **Procedure:** We will consider two methods:
> 1. Graphical: Make an accurate drawing of the three given forces, placing tail to head for the force vectors. The fourth force must bring you back to to the tail of the first force since the system is in equilibrium.
> 2. By components: Calculate the components of each of the three forces and add like components. The fourth force must have components equal to the resultant of the first three and with opposite sign.

The term coplanar means all the forces act in a plane, so the problem is 2-dimensional. We are looking for equilibrium; so there is no information in a motion diagram. Since there is only one body, the only force diagram we can draw is a free body force diagram. This is shown in Figure SG 4.2a. The fourth force has been drawn at an angle ϕ with respect to the reference direction.

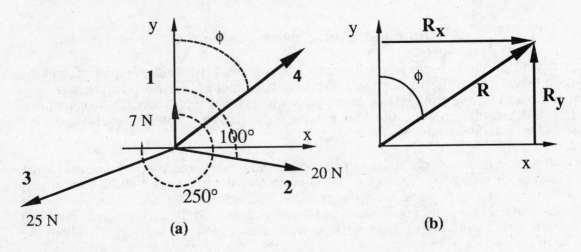

FIGURE SG 4.2:   a) Four forces acting on a 10 kg mass.
                            b) Components of the vector R.

In the figure, we have chosen the reference direction for angle measurement as the y-axis and have also chosen an x-axis. This is done so we can take components to calculate the resultant force of the 3 given forces. That is, for equilibrium,

$$F_1 + F_2 + F_3 + F_4 = 0,$$

where

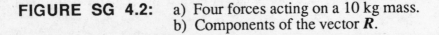

                  $F_1 = 7.00$ N at 0°   ,
                  $F_2 = 20.0$ N at 100° ,
                  $F_3 = 25.0$ N at 250° ,

and $F_4 = R$ N at ϕ°

represents the unknown fourth force with a magnitude of R newtons.

Writing the component equations gives:

$$(F_1 + F_2 + F_3 + F_4)_x = 0$$

and

$$(F_1 + F_2 + F_3 + F_4)_y = 0$$

the components are

$$(F_1)_x = 0, \ (F_1)_y = 7.00 \text{ N}$$

$$(F_2)_x = 20.0 \text{ N} \times \cos(10°) = 20.0 \times 0.985 \text{ N} = 19.7 \text{ N}$$

$$(F_2)_y = 20.0 \text{ N} \times (-\sin(10°)) = -20.0 \times 0.174 \text{ N} = -3.47 \text{ N}$$

$$(F_3)_x = 25.0 \text{ N} \times (-\cos(20°)) = -25.0 \times 0.940 \text{ N} = -23.5 \text{ N}$$

$$(F_3)_y = 25.0 \text{ N} \times (-\sin(20°)) = -25.0 \times 0.342 \text{ N} = -8.55 \text{ N}$$

where we have used Figure SG 4.2a to get the proper angles less than 90° and the proper signs. This is not necessary if you are careful with the sign of the trig functions in each quadrant.

From Figure SG 4.2b we see that for a vector R that $R_x = R \sin \phi$ and $R_y = R \cos \phi$, then for example:

$$(F_2)_x = 20.0 \text{ N} \times \sin(100°) = 20.0 \times (0.985) \text{ N} = 19.7 \text{ N}$$

$$(F_2)_y = 20.0 \text{ N} \times \cos(100°) = 20.0 \times (-0.174) \text{ N} = -3.47 \text{ N}.$$

The signs of the trig functions should be correctly generated by your calculator .

We can now calculate the needed force

$$F_4 = - (F_1 + F_2 + F_3)$$

or

and

$$(F_4)_x = - (0 \text{ N} + 19.7 \text{ N} - 23.5 \text{ N}) = 3.80 \text{ N},$$

$$(F_4)_y = - (7.00 \text{ N} - 3.47 \text{ N} - 8.55 \text{ N}) = 5.02 \text{ N}.$$

Figure SG 4.2b also shows us that:

so

$$R = \sqrt{R_x^2 + R_y^2} \quad \text{and} \quad \tan \phi = \frac{R_x}{R_y}$$

and

$$R = \sqrt{(3.80)^2 + (5.02)^2} \text{ N} = 6.30 \text{ N}$$

so

$$\tan \phi = \frac{3.80 \text{ N}}{5.02 \text{ N}} = 0.75697,$$

$$\phi = 37°.1$$

where we need to know that both components of R are positive to take the inverse tangent.

The entire problem could have been done in radians by restating the initial conditions. This is done by dividing the angle in degrees by 2π. Most calculators will compute trig functions with angles in either degrees or radians and can convert degrees to radians and conversely. Just be sure to be consistent in your choice and to be sure that your calculator is using the units that you input.

You should also note that information was given that is not needed. The mass of the body was not used in this solution. Let this be a warning that the given information for a problem may be more than you need, so don't insist on using all information. This is a better situation than the real world which may let you leave the laboratory without all the needed information.

PROBLEM SG 4.2: A 20 kg block is placed on a frictionless table and connected to a 3 kg block by a string which extends horizontally across the table over a pulley and down to the 3.0 kg block. What is the acceleration of the blocks? Ignore friction in the pulley.

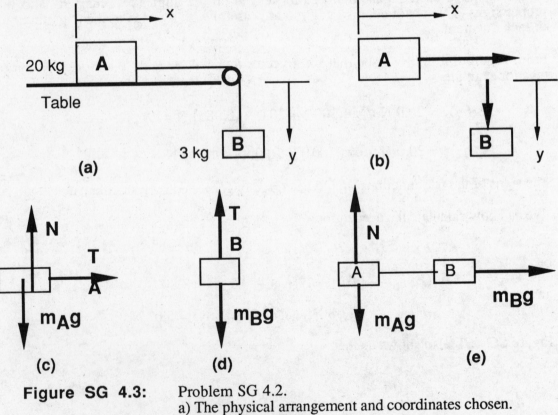

Figure SG 4.3:   Problem SG 4.2.
a) The physical arrangement and coordinates chosen.
b) The motion diagram.
c) Forces on body A.
d) The forces on body B.
e) Forces on the system of both bodies.

SOLUTION:

Procedure: Write the forces. Since the system is frictionless, the only external force which produces motion is on the hanging mass. The gravitational force on the mass on the table is balanced by the normal force. If free body force diagrams are used include the tension of the string on both bodies.

We first draw a general figure showing the problem. This is shown in Figure SG 4.3a. We will choose coordinates x and y, where x represents the path along which the 20 kg block moves and y the direction that the 3.0 kg block moves. We guess the direction of x and y to be the way the blocks move. If we are wrong, we will get negative velocities.

Known Quantities	Symbol
20 kg block block	A
mass of the 20 kg block	m_A
3.0 kg block	B
mass of 3.0 kg block	m_B

We also draw a motion diagram in Figure SG 4.3b. This diagram is good until one of the blocks hits the pulley. The model assumes the pulley and string are ideal.

If we take the free body force diagram approach, we need force diagrams for both bodies. These are shown in Figure SG 4.3c and in 4.3d. Since the only forces on block A are either normal or parallel to the surface, we know that

$$N = -m\,g,$$

so no acceleration occurs to A perpendicular to the table. The remaining equations for A and B are:

$$T_A = m_A a_A$$

and

$$m_B g - T_B = m_B a_B.$$

Since the string is of fixed length,

$$a_A = a_B,$$

and since it is ideal

$$T_A = T_B$$

(Note: there is no minus sign since the direction has already been taken into account).

Applying these two conditions to the preceding equations, we get

$$m_B g = (m_B + m_A)\, a_B$$

or

$$a_B = \frac{m_B}{m_B + m_A}\, g = a_A.$$

If we had chosen to consider the system as a unit, we have the forces shown in Figure SG 4.3e. Again

$$N = -mg,$$

and the acceleration is given by

$$(m_A + m_B)\, a = m_B g$$

which gives the same result.

PROBLEM SG 4.3: A 5.0 kg weight is suspended with a rope (string) and pulley as shown in Figure SG 4.4a. What is the magnitude of the force needed for equilibrium? The pulley has a mass of 0.50 kg.

SOLUTION:

> **Procedure:** Analyze the forces acting on the system. Use components, preferably vertical and horizontal, and require that the sum of the forces is zero. If you use free body diagrams, include tension where appropriate.

We will let F be the magnitude of the force. We idealize the rope as a massless inextensible string. Thus, the force desired must be in the same direction the string points, 30°, and directed outward from the end of the string. For this problem, we must deal with the mass of the pulley since it is given; but, other than that, the pulley is ideal in the sense that it only changes the direction of the string. The connection between the pulley and the 5.0 kg mass is assumed to be an ideal string.

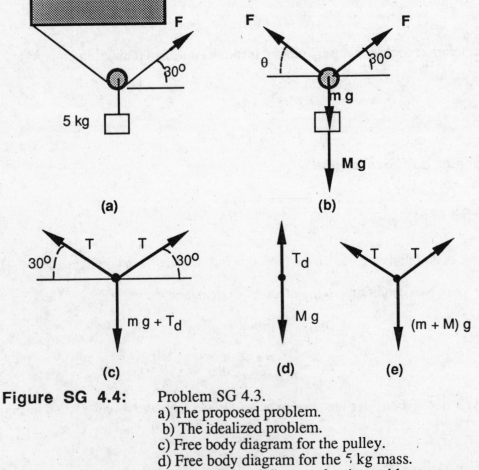

Figure SG 4.4:   Problem SG 4.3.
a) The proposed problem.
b) The idealized problem.
c) Free body diagram for the pulley.
d) Free body diagram for the 5 kg mass.
e) External force diagram for the problem.

Figure SG 4.4b shows the idealized problem and the external forces acting on the system. We will measure vertical directions + downward from the supporting surface.

GIVEN or to be FOUND:

Quantity	Symbol
5.0 kg mass	M
0.50 kg mass	m
tension in the string supporting the pulley	T
tension in the string between the pulley and the 5 kg mass	T_d
applied force	F
angle between left support and horizontal	θ

Since the string holding the pulley is ideal, we know the tension in the string is equal in magnitude to the applied force F. Since the string is ideal, the tension in the string on the right hand side of the pulley has to be the same tension as on the left hand side. Since we are in equilibrium the horizontal forces must add to zero and hence

$$T \cos \theta = T \cos 30° \qquad \text{so} \qquad \theta = 30°.$$

The free body force diagrams for the pulley and the mass are shown in Figure SG 4.4c and 4.4d. We know the horizontal forces cancel since the system is in equilibrium. So only the vertical components need be considered. The equations of equilibrium for the pulley are:

$$-T \sin \theta - T \sin \theta + m g + T_d = 0,$$

while for the mass:

$$-T_d + M g = 0.$$

The unknown force is equal in magnitude to the tension, T, and directed at an angle of 30° along the string. Solving the equations we have:

$$2 T \sin 30° = m g + T_d$$

and

$$M g = T_d.$$

This gives

$$T = \frac{(m + M) g}{2 \sin \theta} = \frac{5.5 \text{ kg} \times 9.8 \text{ m/s}^2}{2 \times 0.5} = 54 \text{ N} = F.$$

The force diagram for the system of pulley plus mass is shown in Figure SG 4.4e. For equilibrium the vertical forces add as

$$-T \sin \theta - T \sin \theta + (m + M) g = 0,$$

which gives, as it must, the same result.

PROBLEM SG 4.4: What is the minimum force required to drag a heavy carton across the floor if the force is applied upward at an angle of 40° to the horizontal? Let the mass of the carton be 85 kg and the coefficient of friction be 0.75.

SOLUTION:

> **Procedure:** Write the forces on the box in component notation. Since the box stays on the floor, the net vertical component must be zero. Use this to find the normal force and substitute into the horizontal component equation.

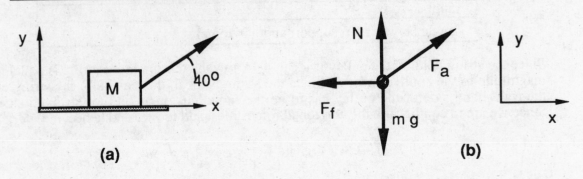

Figure 4.5: Problem SG 4.4.
a) The physical arrangement. b) Force diagram.

This is posed as a problem in dynamic equilibrium; that is, the assumption is that the force is the minimum needed to maintain a constant velocity of the carton. First draw a diagram showing the physical problem as shown in Figure SG 4.5a and a free body force diagram as in Figure SG 4.5b. We expect the forces acting to be gravity, the normal force of the floor, friction, and the applied force.

We denote the applied force by F_a, the mass of the carton by m, The normal force of the floor on the carton by N, and the frictional force by F_f. We choose coordinates x positive to the right and y positive upwards. The conditions for equilibrium are:

Sum of x-components of forces = 0

Sum of y-components of forces = 0.

Using the free body diagram Figure SG 4.5b we have

x-force components

$$F_a \cos \theta - F_f = 0,$$

y-force components

$$F_a \sin \theta + N - m g = 0$$

where we have made use of our knowledge that N must point in the y-direction and F_f must point in the negative x-direction.

GIVEN:

$$m = 85 \text{ kg}$$
$$\theta = 40°$$
$$F_f = \mu N \qquad \text{where } \mu = 0.75$$

Note: We can not assume $N = mg$ since there is a component of F_a in the y direction.

We can put the relation for F_f into the x-component equation to get,

$$F_a \cos \theta = \mu N.$$

The the y-component becomes

$$F_a \sin \theta + \frac{F_a \cos \theta}{\mu} - m g = 0$$

or

$$F_a = \frac{m g}{\sin \theta + \dfrac{\cos \theta}{\mu}} = \frac{833 \text{ N}}{.643 + \dfrac{.766}{.75}} = 501 \text{ N} = 5.0 \times 10^2 \text{ N}.$$

The applied force must be 5.0×10^2 N.

Note: a value of 501 N ignores significant figures and a value of 500 N implies more significant figures than are known.

PROBLEM SG 4.5: What is the minimum force required to push a heavy carton across the floor if the force is applied downward at an angle of 40° to the horizontal? Let the mass of the carton be 85 kg and the coefficient of friction be 0.75.

SOLUTION:

Procedure: Use the same procedure as was used for the preceding problem.

This is identical with problem SG 4.4 except for the direction of the applied force. We will carry the previous analysis through step-by-step with the same detail to show how the analysis is identical except for some sign changes. The numerical results will be very different.

This is a problem in dynamic equilibrium; that is, the assumption is that the force is the minimum needed to maintain a constant velocity of the carton. First draw a diagram showing the physical problem as shown in Figure SG 4.6a and a free body force diagram as in Figure SG 4.6b. We expect the forces acting to be gravity, the normal force of the floor, friction and the applied force.

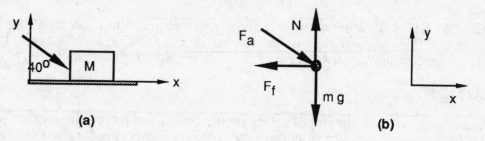

Figure 4.6: Problem SG 4.5. a) The physical arrangement. b) Force diagram.

We denote the applied force by F_a, the mass of the carton by m, The normal force of the floor on the carton by N, and the frictional force by F_f. We choose coordinates x positive to the right and y positive upwards. The conditions for equilibrium are:

Sum of the x-components of forces $= 0$

Sum of the y-components of forces $= 0$.

Using the free body diagram Figure SG 4.6b, we have

x-force components $\qquad\qquad F_a \cos \theta - F_f = 0,$

y-force components $\qquad\qquad -F_a \sin \theta + N - m g = 0$

where we have made use of our knowledge that N must point in the y direction and F_f must point in the $-x$ direction.

$$\textbf{GIVEN: } m = 85 \text{ kg} \qquad \theta = 40° \qquad F_f = \mu N$$

where $\mu = 0.75$ is the coefficient of friction.

also note that we can not assume $N = mg$ since there is a component of F_a in the y direction.

We can put the relation for F_f into the x-component equation to get,

$$F_a \cos \theta = \mu N.$$

The the y component becomes

$$-F_a \sin \theta + \frac{F_a \cos \theta}{\mu} - m g = 0$$

or

$$F_a = \frac{m g}{\dfrac{\cos \theta}{\mu} - \sin \theta} = \frac{833 \text{ n}}{\dfrac{.766}{.75} - 0.643} = 2.2 \times 10^3 \text{ N}.$$

This is a much larger force than we found in Problem SG 4.5. Clearly it is better to pull up on the box to move it, than to push down.

PROBLEM SG 4.6: What is the minimum force required to push an 80 kg carton up an incline of 25° if the coefficient of friction between the incline and the box is 0.70 and if the force is applied upward at an angle of 20° with respect to the horizontal.

SOLUTION:

Procedure: Analyze the forces acting on the box. Since there is no motion perpendicular to the plane, a choice of coordinates parallel and perpendicular to the plane will simplify the solution. Then a component expression has no motion and hence no acceleration perpendicular to the plane. The perpendicular component can be used to eliminate the normal force. The parallel component is then analyzed to determine the minimum size force needed for the acceleration to be zero or directed up the plane.

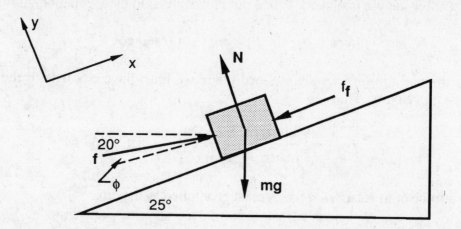

Figure SG 4.7: Problem SG 4.6

NOTE: The problem posed is sketched in Figure SG 4.7. A choice of coordinates has been made with the axes parallel and perpendicular to the plane. This choice is not essential but it has the good feature that any motion occurs along only one of the coordinate axes.

A choice of vertical and horizontal axes would have involved motion in both directions and we would have had to determine that the motion was confined to the surface of the inclined plane. To determine that the motion was on the plane, we would have had to require that for any motion Δ(horizontal) and Δ(vertical) that

$$\frac{\Delta(\text{vertical})}{\Delta(\text{horizontal})} = \tan(25°).$$

This would make a relatively easy problem appear very hard.

ADVICE: PICK YOUR COORDINATE AXES SO AS TO MINIMIZE THE NUMBER OF COORDINATES THAT CHANGE.

Since the force is given as at an angle of 20° upwards with respect to the horizontal, it will be downwards at an angle of $20° - 25° = -5°$ (the minus sign tells us that it is downwards) with respect to the surface of the inclined plane. We will call this angle ϕ. We will also let f represent the magnitude of the applied force and N represent the magnitude of the normal force. The inclination of the plane will be represented by θ.

Newton's second law then gives us

sum of forces in y-direction $= 0 = N - mg \cos \theta - f \sin \phi$
and
sum of forces in x-direction $= ma = f \cos \phi - mg \sin \theta - \mu N$

where a is the acceleration in the x-direction and μ is the coefficient of friction.

First solve the y-equation for N and put the results into the x-equation giving:

$$ma = f[\cos\phi - \mu\sin\phi] - mg[\sin\theta + \mu\cos\theta].$$

For the box to move up the inclined plane we must have $a \geq 0$ with the $= 0$ giving constant velocity. Thus

$$f \geq mg\frac{\sin(\theta) + \mu\cos(\theta)}{\cos(\phi) - \mu\sin(\phi)} = 8.1 \times 10^2 \text{ N}.$$

So a force of at least 8.1×10^2 N must be applied to the box.

PROBLEM SG 4.7: A classroom demonstration is done with an Atwood machine. The mass on the right is 1.25 kg. This mass descends a distance of 2.0 m in 3.2 s from rest. What is the other mass? Ignore effects of pulley mass and friction.

SOLUTION:

Procedure: To choose our method of solution for this problem we need to analyze the given information. We know from the text that there is a simple relation between the masses and the acceleration of the mass pair. We know the system starts at rest and moves a known distance is a known time. This is sufficient information to determine the acceleration. We then can find the mass.

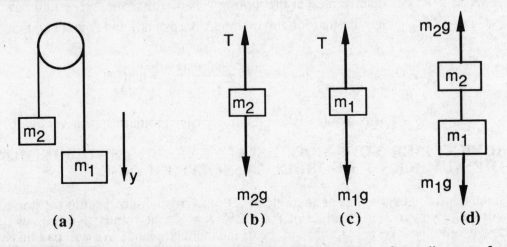

Figure SG 4.8:   a) The Atwood machine, b) and c) Free force diagrams for the two masses, d) net forces on the system.

An Atwood machine is shown in Figure SG 4.8a. We have chosen the direction of motion to be positive downward for both masses. Since we are told that the mass on the right descends, The mass on the right must be greater than the mass on the left. We use the following symbols:

GIVEN

quantity	symbol	value
mass on right	m_1	1.25 kg
mass on left	m_2	?
acceleration of gravity	g	9.8 m/s^2
acceleration of mass m_i	a_i	?(but equal and opposite)
tension in string	T	?

The forces acting on m_2 are shown in Figure SG 4.8b and the forces on m_1 are shown in Figure SG 4.8c. Newton's second law (Σ forces = m a) gives:

$$m_2\, a_2 \;=\; m_2\, g - T$$

$$m_1\, a_1 \;=\; m_1\, g - T,$$

combining gives

$$(m_1 + m_2)\, a_1 \;=\; (m_1 - m_2)\, g$$

where use has been made of the constraint,

$$a_1 = - a_2.$$

A constraint is any relation that restricts the possible values that a variable may have.

As an alternative approach the problem can be considered as shown in Figure SG 4.8d. For this method the motion in the y-direction must be taken as positive for both masses. The force equation is:

$$(m_1 + m_2)\, a_1 \;=\; m_1\, g - m_2\, g,$$

which is the same equation as before. Solving the equation gives:

$$m_2 = m_1 \frac{g - a_1}{g + a_1}\,.$$

To find m_2, we need to know a_1. To get this, use the kinematic law for constant acceleration:

$$y \;=\; y_0 + \Delta y \;=\; y_0 + v_0\, \Delta t + \frac{1}{2}\, a\, (\Delta t)^2,$$

where for this problem we have $v_0 = 0$.

This relation gives the acceleration as:

$$a = \frac{2\, \Delta y}{(\Delta t)^2} \;=\; \frac{2 \times 2.0 \text{ m}}{(3.2 \text{ s})^2} \;=\; 0.39 \text{ m/s}^2,$$

hence

$$m_2 \;=\; 1.25 \text{ kg} \times \frac{(9.80 - 0.39)\text{m/s}^2}{(9.80 + 0.39)\text{m/s}^2} \;=\; 1.2 \text{ kg. }.$$

The mass on the left is 1.2 kg.

PROBLEM SG 4.8: A body of mass 20.0 kg moves with constant velocity of 10 m/s on a horizontal surface under the action of a horizontal force of 15 N. What is the coefficient of friction for the surface and how long will it take the body to stop if the applied force is removed?

SOLUTION:

Procedure: Draw a free body diagram. There is no vertical force so the vertical forces sum to zero. The sum of the horizontal forces will give the acceleration of the body along the surface.

The forces are shown in Figure SG 4.9. The mass of the body is m, the acceleration of gravity is g, the horizontal acceleration of the body is a, the normal force is N, the applied force is f, and the frictional force is $F_f = \mu N$.

(Vertical forces): $$N - mg = 0$$

(Horizontal forces) $$F - \mu N = ma$$

Solving these for the first part when $a = 0$ gives,

$$\mu = \frac{F}{mg}.$$

If v_0 is the initial velocity and v is the velocity at a later time, the relation for constant acceleration is

$$v = v_0 + at.$$

Since the applied force is now zero, the acceleration is given by

$$a = -\frac{\mu N}{m}$$

To find the time it takes the body to stop, set $v = 0$ and solve for t, getting

$$t = \frac{v_0}{-a} = \frac{mv_0}{\mu N}.$$

Entering the numbers gives us $\mu = 0.077$ and the stopping time is about 13 s.

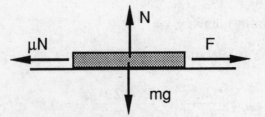

Figure SG 4.9:   Free body force diagram.

Chapter 5
Uniform Circular Motion and Gravitation

Important Terms

Fill in the blank with the appropriate word or words.

1. The _____ is the constant which appears in Newton's Law of Gravitation.

2. The quantity 2π divided by the period of rotation of an object is called the object's _____.

3. The force needed to keep a body in a circular orbit is called the _____.

4. The SI unit of frequency is the _____.

5. The time it takes to complete one revolution about any closed path is called the _____ of the motion.

6. Newton proposed the _____ which tells us the gravitational force between two bodies.

7. The _____ of a substance is the mass per unit volume.

8. The reciprocal of the period of motion is the _____.

9. Motion in a circular path with constant speed is called _____ .

10. The technical name for the acceleration needed to move in a circle at constant speed is the _____ acceleration.

11. The _____ for circular motion is the rate of change of the angular position as seen from the center of the orbit.

12. The _____ is the ratio of the gravitational force on a body to the mass of the body.

13. Drawing _____ of force can help us visualize the gravitational field.

Write the definitions of the following:

14. angular frequency
(both concept and equation with symbols defined) (5.1):

15. hertz (5.1):

16. gravitational field strength
(both concept and equation with symbols defined) (5.5):

17. uniform circular motion (5.1):

18. law of universal gravitation
    (both concept and equation with symbols defined) (5.3):

19. universal gravitational constant
    (both concept and value with units) (5.3):

20. frequency (5.1):

21. angular velocity (5.1):

22. centripetal acceleration (5.1):

23. centripetal force (5.2):

24. density (5.3):

25. period (5.1):

26. lines of force (5.5):

Answers to 1–13, Important Terms

1. gravitational constant; 2. angular frequency; 3. centripetal force; 4. hertz; 5. period;
6. law of universal gravitation; 7. density; 8. frequency; 9. uniform circular motion;
10. centripetal; 11. angular velocity; 12. gravitational field strength; 13. lines.

Comments on Centripetal Force

When labeling a force diagram for an object moving on a circular path, it is best never to label a force only as the "centripetal force". All forces drawn must have a physical existence independent of the particular path the moving body takes. Forces such as that of a rope, the normal force, friction and gravity act on the body. If we are told that the body is moving on a circular path, then we can identify one or more of the physical forces as those causing the circular motion. We can then set the sum of these forces equal to the value of the centripetal force needed for circular motion. For example, consider Figure 5.8 of the text. The automobile is moving on a circle whose center lies to the right. The car's velocity changes direction, so we know there is a centripetal force. The only physical force that can provide this force is the frictional force $\mathbf{F}_{fr}$ between the tires and the road. Thus the total frictional force (that is, the force provided by the 4 tires) in the radial direction must be identified as the centripetal force. So:

$$\sum_{4\ tires}(\mathbf{F}_{fr})_{radial} = \frac{m\,v^2}{r}$$

where m is the mass of the auto, v the speed of the auto and r is the radius of the turning circle. This allows us to calculate the frictional forces that must exist for the motion to be circular.

What may be unclear to you is how we knew that frictional forces existed. Consider what happens when you try to turn a car on glare (smooth) ice. When you try to change direction by turning the steering wheel, you may start to spin but you continue in a straight line. On glare ice there is little or no friction between the tires and the ice. On a normal roadway, changing the steering causes the direction of motion of the car to change. This tells us that there must be friction between the tires and the road for a car to turn.

When analyzing the physical forces, always choose a frame of reference that is not rotating or at least rotates only a small amount during the time under consideration. For this course, unless you are explicitly told otherwise for a particular problem, neglect the rotation of the Earth. When you can not neglect the rotation of your reference frame, there will be other coordinate dependent forces, such as the Coriolis force.

Binomial Expansion

We will often have expression of the form

$$(1 + x)^n$$

where $x << 1$ and n is any number positive or negative.
If n is a positive integer we know that

$$(1 + x)^1 = 1 + x$$

$$(1 + x)^2 = 1 + 2x + x^2$$

$$(1 + x)^3 = 1 + 3x + 3x^2 + x^3$$

or in general

$$(1 + x)^n = 1 + \frac{nx}{1} + \frac{n(n-1)x^2}{1(2)} + \frac{n(n-1)(n-2)x^3}{1(2)(3)} + \dots + x^n$$

This is an increasing power series in x. If $x << 1$, then each successive term is smaller than the preceding, so we can approximate the series as

$$(1 + x)^n \approx 1 + nx.$$

This approximation works even if n is not a positive integer provided $x << 1$.

Sample Solutions

PROBLEM SG 5.1: A merry-go-round (carrousel) has a diameter of 50 feet and a period of 25 s. What is its frequency in hertz? A ticket taker hangs on an outer pole. What is her angular velocity in rad/s, her speed in m/s and her radial acceleration in m/s^2? A rider is 15 feet from the center. What is his angular velocity, speed and radial acceleration in the same units. Verify that both have the same angular velocity using Equation 5.3.

SOLUTION:

> **Procedure:** For this problem we are given a period of rotation and several different radii. This suggests using the relation between angular and linear velocity and the formula for centripetal acceleration. We choose to use meters and radians as our units.

We first draw a sketch showing the idealized problem. The ticket taker is idealized as a point mass 25 feet from the center and the rider as a point 15 feet from the center. The merry-go-round is a disk rotating with a frequency ω which corresponds to a 30 s period. Numbers have apparently been given to 2 significant figures.

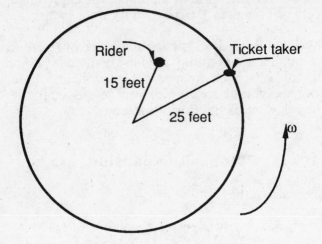

Figure SG 5.1   Merry-go-round with rider and ticket taker.

The frequency, f, is given by text Equation 5.5 as

$$f = \frac{1}{T}$$

where T is the period, so

$$f = \frac{1}{25 \text{ s}} = \frac{0.040}{\text{s}} = 0.040 \text{ hertz.}$$

The ticket taker is at a distance of

$$25 \text{ feet} = 25 \text{ feet} \times \frac{12 \text{ in}}{\text{foot}} \times \frac{0.0254 \text{ m}}{\text{in}} = 7.62 \text{ m}$$

from the center of rotation. Her angular frequency is then

$$\omega = 2\pi \times f = 0.2513272 \text{ rad/s} = 0.25 \text{ rad/s.}$$

The choice of radians as the unit of angular frequency makes the conversion from frequency to angular frequency the factor of 2π.

The circumference of a circle of radius $r = 7.62$ m is
$$2\pi r = 2\pi \times 7.62 \text{ m} = 47.8778316 \text{ m} = 48 \text{ m}$$

to 2 significant figures. Thus the speed of the ticket taker, v, is given by

$$v = \frac{\text{distance}}{\text{time}} = \frac{48 \text{ m}}{25 \text{ s}} = 1.92 \text{ m/s} = 1.9 \text{ m/s.}$$

This means the ticket taker has a centripetal acceleration, $(a_c)_{tt}$, given by

$$(a_c)_{tt} = \frac{v^2}{r} = \frac{(1.9 \text{ m/s})^2}{7.6 \text{ m}} = .475 \text{ m/s}^2 = 0.48 \text{ m/s}^2.$$

The rider is 4.572 m = 4.6 m from the center, so he is moving on a circle of radius 29 m. His speed is then 1.2 m/s and he has a centripetal acceleration, $(a_c)_r$, given by

$$(a_c)_r = \frac{v^2}{r} = \frac{(1.2 \text{ m/s})^2}{4.6 \text{ s}} = 0.31 \text{ m/s}^2.$$

The angular frequency, ω_{tt}, of the ticket taker is

$$\omega_{tt} = \frac{v}{r} = \frac{1.9 \text{ m/s}}{7.6 \text{ m}} = 0.25 \text{ Hz}$$

while the angular frequency, ω_r, of the rider is

$$\omega_r = \frac{1.2 \text{ m/s}}{4.6 \text{ m}} = 0.26 \text{ Hz}.$$

If numbers differ only by one in the last significant figure, they must be considered equal to the given significance. With 2 significant figures these numbers (0.25 Hz and 0.26 Hz) are effectively equal.

PROBLEM SG 5.2: A belt passes over two pulleys. The diameter of the smaller pulley is 6.0 cm while the diameter of the larger is 24 cm. The centers of the pulleys are separated by 2.0 m. If the smaller pulley is turning at a frequency of 100 hertz, what is the frequency of the larger pulley?

SOLUTION:

Procedure: Since the pulleys are connected by a belt, the linear distance a point on one pulley travels must be the same as the distance a point on the other pulley travels.

We let A indicate the smaller pulley and B the larger one. A sketch of the problem is shown in Figure SG 5.2.

The smaller pulley has a circumference given by:

$$\text{small pulley circumference} = \pi \times (6.0 \text{ cm}) = 18.84954 \text{ cm} = 19 \text{ cm}$$

while the larger pulley has a circumference given by:

$$\text{large pulley circumference} = 75 \text{ cm}.$$

When the smaller pulley makes one revolution, the belt moves 19 cm. When the belt moves 19 cm, the larger pulley turns through an angle θ given by

$$\theta = \frac{\text{distance}}{\text{radius}} = \frac{19 \text{ cm}}{12 \text{ cm}} = 1.5833333 \text{ rad} = 1.6 \text{ rad}.$$

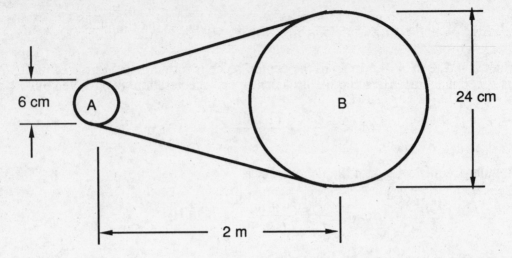

Figure SG 5.2:   Two pulleys connected by a belt.

This angle is $1.6/2\pi = 0.245$ of a complete revolution. The small pulley makes a complete revolution in

$$\frac{1}{f}\,s = 0.01\ s,$$

so the larger pulley makes 0.25 of a turn in 0.01 s. Thus, it will take

$$\frac{0.01\ s}{0.25} = 0.040\ s$$

for a complete revolution. The frequency of the larger pulley is then

$$\frac{1}{0.040\ s} = 25\ Hz.$$

Another approach is to note that the ratio of the diameters of the pulleys is

$$\frac{24\ cm}{6\ cm} = 4.0.$$

This is also the ratio of the circumferences. The belt constrains the system such that when the small pulley makes 4 revolutions, the large pulley must make only one revolution. So the large pulley moves only 1/4 as fast which is 25 Hz.

PROBLEM SG 5.3: What is the breaking strength of a string on which a 0.40 kg stone can be whirled in a horizontal circle with a radius of 1.5 m at a maximum of 2.0 revolutions per second before breaking? Assume the experiment is done on the earth.

SOLUTION:

> **Procedure:** Analyze the forces acting on the stone. These are gravity and the string tension. The vertical component of the tension must equal the gravitational force and the horizontal component must be the same size as the centrifugal force needed for circular motion. The tension in the string must be the maximum tension allowed without breaking the string.

We let the stone be a point of mass m, and the string be an idealized massless string of whose radius of motion is 1.5 m. The frequency f = 2.0 / s. Our problem is to find the breaking strength of the string. The statement of the problem implies that if we had used a more massive rock or if we had spun the rock faster, the string would break. The force which could break the string is the tension in the string. First we make a sketch of the problem which also serves as a motion diagram as shown in Figure SG 5.3a.

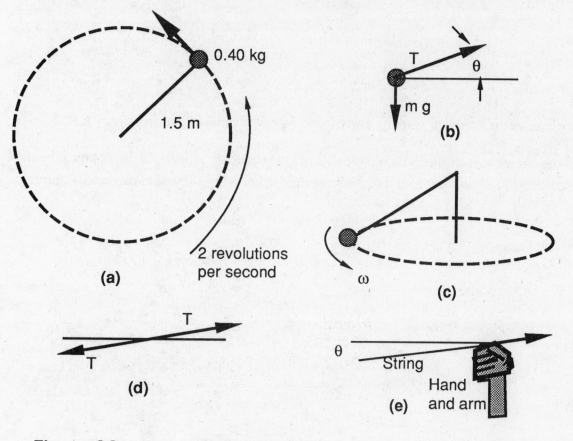

Figure SG 5.3:   a) Preliminary drawing of stone on the end of a whirled string.
b) Forces on the mass.
c) Perspective view of problem.
d) Forces on string.
e) Force on hand from the string

Fig. SG 5.3 shows the forces on the mass. There are only two forces, gravity acting downward and the tension. Clearly for dynamic equilibrium the string can not be exactly horizontal since there must be a vertical component of the tension to balance the force of gravity. The path of the stone can be horizontal as given in the problem, but the string must lie above this path as shown in Figure 5.3c. Other possible free body force diagrams are shown for the string in Figure SG 5.3d and the hand holding the string in Fig. SG 5.3e.

Let the angle between the horizontal and the string be θ. The vertical component of the tension must balance the force of gravity on the stone and the horizontal component of the tension must provide the centripetal force on the stone. The centripetal force is given by Equation 5.6.

GIVEN:

radius of motion	r	1.5 m
frequency	f	2.0 hertz
mass on end of string	m	0.40 kg
tension in string	T	?

The force equations are:

$$m\,g - T \sin \theta = 0 \qquad \text{vertical components}$$

$$T \cos \theta = m\frac{v^2}{r} \qquad \text{horizontal force,}$$

where v is the circular velocity corresponding to the radius r and frequency f.

The vector sum of the two components of the tension $[\ T \sin(\theta), T \cos(\theta)\ ]$ must be the tension, T. So we can find the tension without knowing θ by squaring and summing, that is,

$$T^2 = (T_{\text{horizontal}})^2 + (T_{\text{vertical}})^2$$

or

$$T^2 = T^2 \cos^2 \theta + T^2 \sin^2 \theta = (m\,g)^2 + m^2 \frac{v^4}{r^2}\ .$$

We will need the velocity,

$$v = \frac{\text{circumfrence}}{\text{period}}\ ,$$

$$v = \frac{2 \times \pi \times 1.5 \text{ m}}{\dfrac{1}{2.0/\text{s}}} = 18.85 \text{ m/s}\ .$$

Then

$$T = m\sqrt{g^2 + \frac{v^4}{r^2}} = 0.40 \text{ kg}\sqrt{(96 + 56{,}113) \text{ m}^2/\text{s}^4} = 94.8\text{N}.$$

The tension in the string is 95 N.

PROBLEM SG 5.4: Consider the preceding problem (Problem SG 5.3). The speed is increased slightly and the string breaks. Which of the following sketches (Figure SG 5.4) of motion in the horizontal plane properly shows the subsequent motion in that plane? What will be the motion in the vertical plane?

SOLUTION:

Look at Figure SG 5.4a-c and choose the answer you feel is correct.

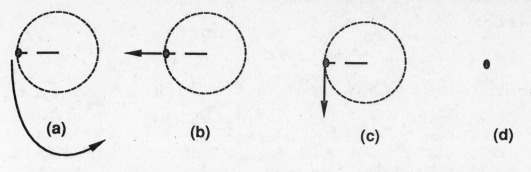

(a)   **(b)**   **(c)**   **(d)**

Figure SG 5.4:   Three possible motions of the stone in the horizontal plane when the string   breaks:
 a) spiral. b) radially outward c) tangential. d) Space for force diagram.

The proper way to answer this question is to draw a force diagram for the stone when the string has broken. Space has been left for you to draw this in Figure SG 5.4d. Please do so before preceding.

The only force is that of gravity acting vertically downward, there are no forces in the horizontal plane. Newton's Second Law then tells us that the velocity must continue with the value it had when the string breaks. Figure (c) must be the correct answer since the velocity was tangential before the string broke.

The vertical motion would correspond to uniform acceleration vertically downward under the force of gravity. The total motion would then be the same as if the stone had been fired horizontally from a cannon with velocity v.

PROBLEM SG 5.5: What would be the fractional change in the weight of a 120 lb woman when she moves from sea level to a mountain 2.0 km high? Assume that she is at the equator.

SOLUTION:

Procedure: Consider what forces might act on the woman. The major force on the woman is gravity; however, the rotation of the Earth will make her appear to weigh less. So we must subtract the centrifugal force from the gravitational force to get her apparent weight.

The only externally applied force we will consider as acting on her is gravity as shown in Figure SG 5.5. Let m be her mass, g_1 the acceleration of gravity at sea level, and g_2 the acceleration of gravity at the top of the mountain.

$m\,g_1$ $m\,g_2$

(a)   **(b)**

Figure SG 5.5:   a) Weight at sea level. b) Weight at 2 km above sea level.

GIVEN:

sea level weight	120 lbs
height change from sea level	$2.0 \text{ km} = 2.0 \times 10^3 \text{ m}$

The fractional change in weight is

$$\frac{(\text{weight at 2 km}) - (\text{weight at sea level})}{\text{weight at sea level}}$$

Now

$$\text{weight} = mg - \frac{mv^2}{r} = mg - m\,\omega^2 r$$

where m is her mass, g the acceleration due to gravity, r the distance from the center of the Earth and ω is the Earth's rotation frequency. If she had not been at the equator, r would be the distance to the Earth's rotation axis rather than the distance to the Earth's center. We let r_1 and r_2 be the respective distances from the Earth's center.

So

$$\text{fractional change} = \frac{(mg_2 - m\,\omega^2 r_2) - (mg_1 - m\,\omega^2 r_1)}{mg_1 - m\,\omega^2 r_1}$$

$$= \frac{g_2 - \omega^2 r_2}{g_1 - \omega^2 r_1} - 1$$

The first thing we note is that the actual weight did not matter when we calculate the fractional change in weight. The acceleration due to gravity we can get from Newton's Second Law and the Universal Law of Gravity:

$$F = mg = \frac{G M m}{r^2}$$

where m is the her mass, M the mass of the Earth and r the distance from the Earth's center. The first equality is from the second law and the second equality is from the gravitational law. Thus we have that

$$g = \frac{G M}{r^2}$$

and the fractional weight change, f, is

$$f = \frac{g_2 - \omega^2 r_2}{g_1 - \omega^2 r_1} - 1 = \frac{GM/r_2^2 - \omega^2 r_2}{GM/r_1^2 - \omega^2 r_1} - 1$$

$$= \frac{\dfrac{r_1^2}{r_2^2} - \dfrac{\omega^2}{GM} r_2 r_1^2}{1 - \dfrac{\omega^2}{GM} r_1^3} - 1 = \frac{\dfrac{r_1^2}{r_2^2} - 1 + \dfrac{\omega^2}{GM} r_1^2 (r_2 - r_1)}{1 - \dfrac{\omega^2}{GM} r_1^3}.$$

The first term

$$\left(\frac{r_1^2}{r_2^2} - 1\right)$$

is the correction due to the change in gravity resulting from the change in distance. The second term,

$$\left(\frac{\omega^2}{GM} r_1^2 (r_2 - r_1)\right),$$

is the correction for the change in the centripetal force. The denominator of the fraction is not one since we used the measured weight at sea level instead of the true weight.

Now, using the values in text examples 5.12 and 5.13:

radius of Earth	$= 6.38 \times 10^6$ m
mass of Earth	$= 5.98 \times 10^{24}$ kg
gravitational constant	$= 6.67 \times 10^{-11}$ N m/kg^2.

Since the Earth rotates with a 24 hr period,

$$\omega = \frac{2\pi}{T} = \frac{2\pi}{24 \text{ hr} \times 3600 \text{ s/hr}} = 7.27 \times 10^{-5} \text{ /s}.$$

We have

$$\frac{\omega^2 r_1^2}{GM} = 5.41 \times 10^{-10}/\text{m},$$

so

$$1 - \frac{\omega^2}{GM} r_1^3 = 1 - 3.45 \times 10^{-3} = 0.997$$

while

$$\frac{\omega^2 r_1^2}{GM} (r_2 - r_1) = 1.08 \times 10^{-6}$$

and if we let $\Delta r = r_2 - r_1$,

$$\frac{r_1^2}{r_2^2} - 1 = \frac{r_1^2}{r_1^2 (1 + \Delta r/r_1)^2} - 1 = (1 + \Delta r/r_1)^{-2} - 1.$$

Now use

$$(1 + x)^n = 1 + nx + \dots \approx 1 + nx \text{ for } x \ll 1$$

and get

$$\frac{r_1^2}{r_2^2} - 1 = -2\frac{\Delta r}{r_1} = -6.3 \times 10^{-4}.$$

We have only two significant figures in the last term and hence the other corrections can be neglected and the fractional change is -6.3×10^{-4}.

We conclude that the centripetal force could have been neglected from the beginning; however, it was necessary to do the complete calculation to know this.

PROBLEM SG 5.6: What is the orbital velocity of an Earth satellite which is in orbit at a radius of 1.02 times the Earth's radius?

SOLUTION:

> **Procedure:** Consider the forces. There is only gravity. Since the orbit is circular, the gravitational acceleration must be equal to the centripetal acceleration.

From the preceding problem the radius of the Earth is 6.38×10^6 m. Make a motion sketch in the space labeled Figure SG 5.6,a where as indicated we take R to be the radius of the Earth, m the mass of the satellite, M the mass of the Earth.

(a)   (b)

Figure SG 5.6:   a) Space to sketch circular orbit of space satellite and forces acting. for Problem SG 5.6.
   b) Space to sketch circular orbit of Titan around Saturn for Problem SG 5.7.

The only force present is the gravitational force given by

$$\text{gravitational force} = \frac{GmM}{(1.02\,R)^2},$$

where M is the mass of the Earth, m is the mass of the satellite and R is the radius of the Earth.

Since the orbit is circular, this force must be the centripetal force

$$\text{centripetal force} = \frac{mv^2}{1.02\,R},$$

where v is the orbital velocity of the satellite.

Equating these gives

$$v^2 = \frac{G\,M}{1.02\,R}$$

$$= \frac{(6.67 \times 10^{-11}\text{N m}^2/\text{kg})(5.98 \times 10^{24}\text{ kg})}{1.02 \times 6.38 \times 10^6\text{ m}} = 6.13 \times 10^7\text{ m}^2/\text{s}^2.$$

So

$$v = 7.8\text{ km/s}.$$

This result is again independent of the mass of the satellite as expected.

PROBLEM SG 5.7: Calculate the mass of Saturn from the orbital data of Titan. Titan orbits Saturn at 1.221×10^6 km with a period of 15.594 days.

SOLUTION:

Procedure: This is the same problem as SG 5.6 except for the numbers used. Follow the same procedure.

First sketch the orbit of Titan about Saturn in the space left in Figure SG 5.6b. This will be a circle with Saturn in the center. Both Saturn and Titan are nearly at the same distance from the sun, so we can treat them is being in the same gravitational field due to the sun. This allows us to ignore the effect of the sun on the motion and only consider the gravitational interactions of the satellite and the planet. It is also reasonable to neglect the motion of Saturn produced by Titan if we can assume that the mass of Saturn is much larger than the mass of Titan. These assumptions allow us to to claim that the gravitational attraction between Saturn and Titan is equal to the centripetal force needed to keep Titan in orbit about Saturn.

GIVEN:

Quantity	Symbol
Mass of Saturn	M
mass of Titan	m
center distance between Saturn and Titan	r
velocity of Titan about Saturn	v
period of Titan about Saturn	T
gravitational force between objects	F_g
centripetal force	F_c

KNOWN:

$$r = 1.221 \times 10^6 \text{ km}, \quad T = 15.594 \text{ days}$$

The force diagram which you have drawn in Figure SG 5.6b gives

$$m\frac{v^2}{r} = F_c = F_g = G\frac{m\,M}{r^2}$$

The velocity can be found from the period by the relation

$$\text{velocity} = \frac{\text{circumfrence}}{\text{period}} = 2\pi\frac{r}{T} = v.$$

Solving these relations will give you the mass of Saturn, M.

PROBLEM SG 5.8: In problem SG 5.3, how long is the length of the string between the rock and the pivot point? Assume the radius of rotation is 1.500 m.

SOLUTION:

> **Procedure:** We have done most of the work in this problem. It is only necessary to determine the angle θ between the plane of rotation and the string. Then simple trig will give us the solution.

Examine Figure SG 5.3. The length of the string is the hypotenuse of a right triangle with long side equal to the radius of the circle of rotation and included angle θ? Thus

$$\cos \theta = \frac{r}{L}$$

where $r = 1.5$ m as given in the problem and L is the desired length.

We found the vertical and horizontal forces to be

$$mg - T \sin \theta = 0$$

and

$$T \cos \theta - m\frac{v^2}{r} = 0$$

so

$$\tan \theta = \frac{\sin \theta}{\cos \theta} = \frac{T \sin \theta}{T \cos \theta} = \frac{mg}{m\frac{v^2}{r}}$$

$$= \frac{r\,g}{v^2} = 0.0407.$$

This gives

$$\theta = 0.041 \text{ rad} = 2.3°.$$

Since this is such a small angle, it is easy to see why many people assume that the pivot point is in the plane of rotation.

The length of the string is then

$$L = \frac{r}{\cos \theta} = \frac{1.500 \text{ m}}{0.999} = 1.502 \text{ m}$$

The radius is only slightly different from the string length. The purpose of this problem is to make sure you understand that these two quantities (radius, length) are different.

Chapter 6
Work and Energy

Important Terms

Fill in the blank with the appropriate word or words.

1. The _____ is the unit of work in the SI system.

2. The time rate of doing work is called _____.

3. _____ is the energy of translational motion.

4. The ability to do work is called _____.

5. The energy that a body has because it is in a gravitational field is called the body's _____ energy.

6. The quantity which plays the same role as mass in the expression for kinetic energy is called the _____.

7. _____ is defined as the product of a force and the distance through which the force acts.

8. The _____ states that the work done on a body is equal to its change in kinetic energy.

9. The _____ is a measure of the energy of rotational motion.

10. A(n) _____ diagram is a plot of the potential energy against the displacement.

11. The unit of power is called the _____.

12. The sum of kinetic and potential energy is called _____ .

13. Our first conservation law is the _____ .

14. Another name for the energy of translational motion is _____ energy.

15. _____ laws tell us which quantities are the same before and after an interaction.

16. For a(n) _____ force, the work done depends upon the starting and ending points and is independent of the path.

17. A quantity which remains constant during a change in a system is said to be _____.

18. A body's _____ energy is the sum of its kinetic energy and its potential energy.

Write the definitions of the following:

19. potential energy diagram (6.7):

20. gravitational potential energy (6.4):

21. work-energy theorem (6.1):

22. total mechanical energy (6.7):

23. watt (and its definition in terms of joule/s) (6.8):

24. instantaneous power (6.8):

25. translational kinetic energy (6.3):

26. rotational kinetic energy (6.6):

27. conservation of mechanical energy (6.7):

28. power (6.8):

29. moment of inertia (6.6):

30. mechanical energy (6.7):

31. kinetic energy (6.3):

32. work (6.1):

33. conservative force (6.7):

34. energy (6.2):

35. conservation laws (6.7):

36. joule (6.1):

27. horsepower (and its definition in terms of the watt) (6.8):

38. distinguish between work and power

Answers to 1-18, Important Terms

1. joule; 2. power; 3. translational kinetic energy; 4. energy; 5. gravitational potential;
6. moment of inertia; 7. work; 8. work-energy theorem; 9. rotational kinetic energy;
10. potential energy; 11. watt; 12. total mechanical energy; 3. conservation of mechanical
energy; 14. kinetic; 15. conservation; 16. conservative; 17. conserved; 18. mechanical

General Comments

With the introduction of rotation we must modify our model of bodies. Instead of points, they must now be extended bodies. There are two models we can take:

1. Deal with the extended body by dividing it into small pieces and sum over the pieces with the appropriate forces. Also add internal forces to keep these pieces in a fixed relationship to each other (rigid body). This approach is used in advanced courses.

2. Treat the body a point mass, but also give a set of numbers which describes the body's rotational properties about different axes. External forces, except for gravity act on the surface of the body.

For this course we will adopt the latter approach: The body has finite size but acts as if all the mass were concentrated at one point of the body (center of mass). The gravitational force will act on the body as if all the mass were at this point. Other forces will act on points of the surface of the body. The body will respond with translation and rotation as given by text Equations 6.5 and 6.9 and change gravitational potential energy as given by Equation 6.6. The moment of inertia is then just another number, like the mass, of the set of numbers that give the properties of a body.

Some moment of inertias are given in Table 6.3 of the text. Notice that the moment depends not only upon the mass and the shape of the body but also upon the location of the axis about which the rotation occurs. It would seem that the moment of inertia is not just characteristic of the body, but also depends upon the axis of rotation. It is possible to define a set of numbers which depend only upon the body (the characteristic moments of inertia for the body) and then give rules allowing one to calculate the moment of inertia for any given axis. Assigned problems will either use the quantities in table 6.3 or will give you the necessary moment of inertia.

Comments on Work and Power

The following questions should always be asked:

Who or what does the work?

Who or what has work done on it?

Is there motion with respect to the force?

If you properly ask and answer these questions you will have few difficulties in expressing the physics of a problem. Failing to ask these questions will make the problems difficult. In the sample solutions we will show examples of how to properly apply these questions.

The work done by a variable force can be found by adding up the area under a force versus displacement curve. This follows from the fundamental definition of Work = Force $\times$ displacement. For example, Figure SG 6.1a shows a force versus displacement plot for a force which goes like the gravitational force $F \propto 1/r^2$. The problem is how to calculate the total work done while moving from one value of r to another. In Figure SG 6.1b we show an enlarged portion of the curve of Figure SG 6.1a between two values r_1 and r_2. We let $F(r)$ be the force at r and let the separation between r_1 and r_2 be small enough that the straight line between $F(r_1)$ and $F(r_2)$ is a good approximation to the curve $F(r)$. Then the area under the curve is approximately the sum of the rectangle and the triangle:

$$\text{area} = (r_2 - r_1) \times F(r_2) + \frac{1}{2} \times (r_2 - r_1) \times [F(r_1) - F(r_2)]$$

$$= (r_2 - r_1) \times \frac{F(r_1) + F(r_2)}{2}$$

$$= \text{displacement} \times \text{average force} = \text{work}.$$

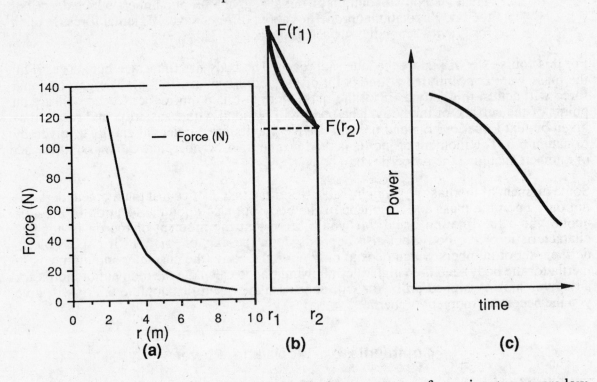

Figure SG 6.1:   a) Force versus displacement curve for an inverse square law.
b). A small section of the curve in a).
c). A power versus time curve.

This procedure can be done for a larger displacement by taking the total displacement as a sum of small displacements. Then the area under a work versus displacement curve is the total work done. The same arguments can be made for a power versus time curve such as shown in Figure SG 6.1c to allow us to conclude that the total work done is the area under the power versus time curve.

When analyzing curves such as in Figure SG 6.1, if a positive displacement indicates work done on the body then a negative displacement would indicate work done by the body.

Sample Solutions

PROBLEM SG 6.1: How much work is done by a man who weighs 550.0 N and who carries a box whose mass is 3.0 kg up a 20.0 m hill? Hint: assume the motion is done with constant velocity.

SOLUTION:

> **Procedure:**
> 1. Force method: Examine the forces. Determine which are exerted **by** the man rather than **on** the man. Choose the component of force in the direction of motion and calculate the work.
> 2. Work–energy method. The work must be the change in potential energy.

1. Force method:

A sketch of the problem is given in Figure SG 6.2a. We let M be the mass of the man, m the mass of the box, g be the acceleration of gravity and h the height of the hill.

If we only had to worry about the work done on the box, we would argue that he applies an upward force of magnitude mg on the box. This force acts directly upwards for a distance h, so the work done by the man on the box is $m\,g\,h$.

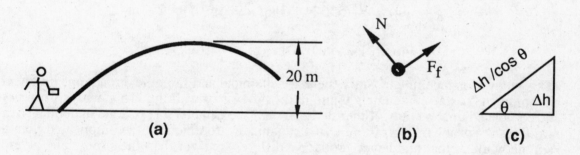

(a)	(b)	(c)

Figure SG 6.2:   a). Sketch of man and box moving up hill.
b). Forces involved..
c). how height moved is related to distance traveled.

The analysis of the work done by the man to move up the hill is much more complicated. The problem is that the object being moved is, in some sense, the source of the work. Let us analyze the problem by examining the forces acting on the man as shown in Figure SG 6.2b. Gravity acts downward on the man box system with a force of

$$\text{Total Gravitational force} = (M + m)\,g$$

and the plane also provides a force which balances out the force. This force by the plane can be broken into two components, a tangential force, F_f, and a normal force, N. The normal force is directed upwards from the plane and balances the downward gravitational force, That is

$$N = mg \cos \theta$$

where θ is the slope of the hill at that location. There is also a frictional force directed up the hill which opposes the downward directed tangential gravitational force. Since the velocity is constant we must have

$$F_f = mg \sin \theta.$$

This frictional force is produced by the interaction of the man and the hill. In order to walk up the hill, the man must contract and release his muscles. The net effect of this is an effective force of

$$\text{force} = (M + m)g \cos \theta$$

where θ is the slope angle. The distance moved as seen in Figure SG 6.2c is just

$$\text{distance} = \frac{\Delta h}{\cos \theta}$$

where Δh is the vertical height traveled at the angle θ. Adding these together and summing over the path gives

$$\text{Work} = \text{force} \times \text{distance} = (M + m)g \cos \theta \times \frac{\Delta h}{\cos \theta} = (M + m)g \, \Delta h$$

$$= (M g + m g) \, h = (550 \text{ N} + 3 \text{ kg} \times 9.8 \text{ m/s}^2) \times 20 \text{ m}.$$

$$= 11588 \text{ N-m} = 1.2 \times 10^4 \text{ N-m}$$

This analysis is incomplete, since there are chemical and biological reactions involved in holding a box at a fixed height and in the process of walking. That analysis can not be completed until a study of thermodynamics (see Chapter 11) is done in depth. Then we find that problems involving living organisms are much more complicated than a simple work picture and hence produce a different answer than the simple one given. However, if we ignore internal changes in the organism, the simple work picture is correct and easily applied provided you use the work-energy approach.

Work-energy method:

Another way to approach this problem is by the work-energy theorem. The difference in energy between the top and the bottom of the hill is just the added potential energy $(M + m) \, g \, h$. This energy change can only be provided by the effort of the man, so that must be the work done.

$$\text{work} = (M + m) \, g \, h.$$

Another simplification that results from the use of the work energy theorem is that we need not deal with the details of human motion and work.

Our definition of work (at this point) does not include the internal processes of living organisms.

PROBLEM SG 6.2: A 50 N force is applied to a box at an downward angle of 30° to the horizontal. The box moves with constant velocity against a frictional force of 43.3 N. How much work is done for a travel distance of 20 m? What is the coefficient of friction? If this takes 10 s, what is the power?

SOLUTION:

> **Procedure:** This problem involves sliding friction. Such a force can not be treated with the work-energy theorem, so us the force equations.

The first step is to draw a motion diagram and a free body force diagram. This is done in Figure SG 6.3a for the motion and 6.3b for the forces.
We let

d	distance traveled,
$F_{applied}$	applied force,
$F_{friction}$	frictional force,
m	mass of the box,
θ	30° angle,
g	acceleration due to gravity,
N	normal force
μ	coefficient of friction.

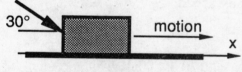

(a)

(b)

Figure SG 6.3:   a) Motion diagram for the problem.
b) Free body force diagram. Compare b) with Figure SG 4.6b. They are the same except for the angle (30° vs 40°)

Since the velocity is constant, the system is in dynamical equilibrium and the sum of the forces is zero.

x-component $\qquad F_{applied} \cos \theta - F_{friction} = 0$

y-component $\qquad F_{applied} \sin \theta + mg - N = 0.$

The motion is in the x-direction, so only the x-component forces need be considered to calculate the work done.
The work done by the applied force is given by

$$\text{work} = F_{applied} \times \cos \theta \times d$$

$$= 50 \text{ N} \times 0.8660254 \times 20 \text{ m} = 8.7 \times 10^2 \text{ J}.$$

Friction affects the amount of work done. If the frictional forces were smaller, a smaller applied force would be needed to keep the box moving at a constant velocity, so less work would be done. In the absence of friction no work would be needed to move the box once it reached the velocity of travel. That is

$$\text{work by friction} = - \text{ work by applied force.}$$

The coefficient of friction is given by

$$F_{\text{friction}} = \mu N.$$

We are given the magnitude of the frictional force, which is not sufficient to find the coefficient of friction. That is, we have two equations, which we can solve for μ as

$$\mu = \frac{F_{\text{applied}} \cos \theta}{F_{\text{applied}} \sin \theta + mg},$$

We will need the mass of the box to find the coefficient of friction. Since this was not given, the coefficient of friction can not be found.

We can get the power. The total work done is

$$\text{work} = F_{\text{applied}} \times \cos(\theta) \times d.$$

This work is done in 10 s so the power is:

$$\text{power} = \text{rate of work} = \frac{\text{work}}{\text{time}} = \frac{870 \text{ J}}{10 \text{ s}} = 87 \text{ W}.$$

Another way to get the power is by

$$\text{power} = \text{force times velocity.}$$

The box moves 20 m in 10 s, so its speed, v, is 2 m/s. Thus, the power is:

$$\text{power} = F \times v = F_{\text{applied}} \cos \theta \times v$$

$$= 50 \text{ N} \times 0.8660254 \times 2 \text{ m/s} = 87 \text{ W}.$$

PROBLEM SG 6.3: 30 J is required to extend a spring 5.0 cm from its equilibrium position. How much force is required to hold the spring there?

SOLUTION:

> **Procedure:** The work done is given. This work should be determined by the force applied and the distance, or since this is a spring we can use the work-energy theorem to relate the work done to the spring constant and displacement. This determines the distance and then by Hooke's Law the force.

First draw a motion diagram as in Figure SG 6.4a. Once the spring is extended there are two forces acting: F_s, the force of the spring to the left and F_a, the applied force to the right. Draw these forces in the space left for Figure SG 6.4b.

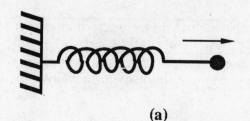

(a)

(b)

Figure SG 6.4:   a). Motion diagram.
b) . Space for force diagram at final equilibrium.

The work done, W, to extend a spring a distance x is given by text Equation 6.3 as

$$W = \frac{1}{2} k x^2$$

where k is the spring constant.

The force, F, needed to keep the spring extended a distance x is

$$F = kx.$$

From Figure SG 6.4b we see that

$$F_a = F_s = k x.$$

GIVEN:

Quantity	Symbol	Value
Work done	W	30 J
displacement	x	5.0 cm.
spring constant	k	?
applied force	F_a	?

We can use the work equation to find k and then use the force equation to find F_a; that is,

$$W = \frac{1}{2} k x^2 \text{ is solved for } k = \frac{2W}{x^2}.$$

This value for k can be substituted into $F_a = k x$ to get

$$F_a = \frac{2W}{x} = \frac{2 \times 30 \text{ J}}{0.050 \text{ m}} = 1.2 \times 10^3 \text{ N}.$$

So it requires 1.2×10^3 N to hold the spring.

PROBLEM SG 6.4: What is the kinetic energy of a 15 gram bullet which is fired at 1200 m/s from a 5 kg rifle? Assume the velocity is given relative to the ground.

SOLUTION:

> **Procedure:** The kinetic energy is given by the mass and the velocity of an object so a straight forward application of the kinetic energy law should suffice.

Kinetic energy is given as $1/2\ m\ v^2$ where m is the mass and v is the velocity. To answer this question we do not need the mass of the rifle or even to know how the bullet reached the speed v.

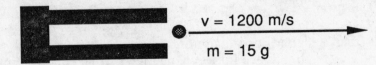

Figure SG 6.5:   Bullet leaving rifle.

GIVEN:

mass	m	15 gram $= 1.5 \times 10^{-2}$ kg
velocity	v	1.200×10^{3} m/s

Then

$$\text{kinetic energy} \quad = \frac{1}{2}\, m\, v^2$$

$$= \frac{1}{2} \times 1.5 \times 1.44 \times 10^{-2+6}\ \text{kg m}^2/\text{s}^2$$

$$= 1.1 \times 10^4\ \text{J}$$

Usually the velocity of a bullet is given relative to the gun (the muzzle velocity). In that case we must also consider the recoil of the gun in order to calculate the velocity of the bullet relative to the ground. This problem can be considered after you complete the next chapter.

PROBLEM SG 6.5: A drum full of hardened cement is rolled down an incline at a warehouse for removal and disposal. The drum has a diameter of 0.50 meter, a length of 1.0 meter and a mass of 750 kg. The incline is at 30° and lowers the drum 1.2 m. What is the translational velocity of the drum just after it leaves the ramp. Assume the drum rolls without slipping and that it was rolling at a translational velocity if 1.0 m/s just as it starts down the ramp.

SOLUTION:

> **Procedure:** Since there is no slipping, we can relate the angle through which the drum has rotated to the distance it has traveled. Also, since no work is done by frictional forces, we can use conservation of energy. Total energy { = kinetic + potential } is a constant of the motion.

This problem is most easily solved by energy conservation. By use of the conservation theorem we can eliminate the need to know all the forces. In Figure SG 6.6a we show the initial conditions and in Figure SG 6.6b the drum just after leaving the ramp. We use the following:

GIVEN:

quantity	symbol used	value given
radius of drum	r	0.25 m
length of drum	b	1.0 m
mass of drum	m	750 kg
initial height	h	1.2 m
incline angle	θ	30°
horizontal distance moved	d	$h/\tan\theta$
initial translational velocity	v_0	1.0 m/s
initial rotational velocity	ω_0	?
final translational velocity	v	?
final rotational velocity	ω	?

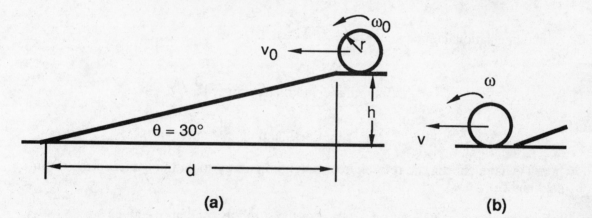

Figure SG 6.6: a). Drum just before it starts down the ramp.
b). Drum just after leaving the ramp.

Since the drum rolls without slipping there is a relation between the angle through which the drum has turned and the distance it has moved:

$$\text{angle turned} = \frac{\text{distance}}{2\pi\ \text{radius}} \times 2\pi.$$

Where the ratio tells us the number of rotations and the last 2π converts this to radians. (2π radians = one circle) Dividing both sides of this relation by time we can convert this to a relation between translational velocity and angular velocity:

$$\text{angular velocity} = \frac{\text{translational velocity}}{r}.$$

This is just Equation 5.3 of the text, which gives

$$\omega_0 = \frac{v_0}{r} \quad \text{and} \quad \omega = \frac{v}{r}.$$

NOW YOU MUST CHOOSE THE ZERO FOR POTENTIAL ENERGY

We choose the zero of potential energy to be the height of the center of the drum after it leaves the ramp. Any other choice is just as good.

The conservation of mechanical energy theorem gives

initial translational kinetic energy $\left(\frac{1}{2} m \, v_0{}^2\right)$

$+$ initial rotational kinetic energy $\left(\frac{1}{2} I \, \omega_0{}^2\right)$

$+$ initial potential energy $(m \, g \, h)$

$=$

final translational kinetic energy $\left(\frac{1}{2} m \, v^2\right)$

$+$ final rotational kinetic energy $\left(\frac{1}{2} I \, \omega^2\right)$

$+$ final potential energy (0)

where we have chosen the reference of potential energy to be the center of the drum's final height.

Writing the energy conservation in terms of the symbols used above gives

$$\frac{1}{2} m \, v_0{}^2 + \frac{1}{2} I \, \omega_0{}^2 + m \, g \, h = \frac{1}{2} m \, v^2 + \frac{1}{2} I \, \omega^2 + 0$$

using the relations between ω and v (and ω_0 and v_0) allows the elimination of the ω giving:

$$\frac{1}{2}\left(m + \frac{I}{r^2}\right)v_0{}^2 + m \, g \, h = \frac{1}{2}\left(m + \frac{I}{r^2}\right)v^2$$

which gives us

$$v^2 = v_0{}^2 + \frac{m \, g \, h \, r^2}{m \, r^2 + I}$$

$$= v_0{}^2 + \frac{gh}{\left(1 + \dfrac{I}{mr^2}\right)}$$

The moment of inertia to use is just that for a disk rotating about one edge parallel to the axis of symmetry which is given in Table 6.3 of the text as

$$I = \frac{3}{2} m r^2.$$

So

$$v^2 = v_0^2 + \frac{2}{5} g h$$

and

$$v = 2.4 \text{ m/s.}$$

ANOTHER CHOICE FOR THE ZERO OF POTENTIAL ENERGY

Had you chosen the reference for potential energy differently, the results would have been the same. For example, suppose we choose the zero of potential as the center of the drum at the top of the ramp, then the conservation of mechanical energy theorem gives us:

initial translational kinetic energy $\left(\frac{1}{2} m \ v_0^2\right)$

+ initial rotational kinetic energy $\left(\frac{1}{2} I \ \omega_0^2\right)$

+ initial potential energy (0)

=

final translational kinetic energy $\left(\frac{1}{2} m \ v^2\right)$

+ final rotational kinetic energy $\left(\frac{1}{2} I \ \omega^2\right)$

+ final potential energy $(-m g h)$.

This gives us the same results.

$$\frac{1}{2} m v_0^2 + \frac{1}{2} I \omega_0^2 + 0 = \frac{1}{2} m v^2 + \frac{1}{2} I \omega^2 - mgh$$

It is also interesting to note that the result is independent of the length and of the mass of the drum. This means that a drum of styrofoam would travel the same as the cement. This is not too surprising since they would fall at the same rate if dropped. The independence of the result on the mass is then a rough check of our results.

PROBLEM SG 6.6: A ball is thrown upwards with sufficient speed to reach a height of 12.0 m, what is its speed when it reaches a height of 10 m ?

SOLUTION:

Procedure:
1. Energy Conservation method: The sum of kinetic energy and potential energy is constant.
2. Kinematical approach. Write the equation for constant acceleration. Use the value of the maximum height to determine constants and then solve for time at 10 m.

There are two ways of approaching this problem: the energy conservation methods of this chapter and the kinematical equations of chapter 2. This problem provides a good comparison of the methods. We will choose h as the vertical coordinate and x as the horizontal coordinate and make a motion diagram sketch as shown in Figure SG 6.7a. The only force acting is gravity, vertically downward as shown in Figure SG 6.7b.

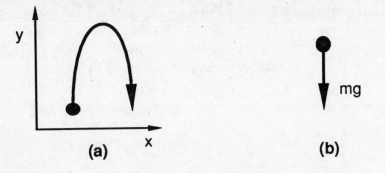

Figure SG 6.7:   a). Motion of ball in a gravitational field.
b). Free body force diagram for ball.

GIVEN:

velocity at 12.0 m	0.0 m/s
maximum height h_0	12 m

1. ENERGY APPROACH

Use the energy conservation theorem:

Kinetic energy + potential energy = constant.

Let us call this constant k.

Since the x-and y-components can be treated separately and since there are no x-forces, we will treat the problem as if only y-motion occurs and represent the y-velocity by v. Picking the potential energy origin at $h = 0$, we have

$$\text{Kinetic energy}\left(\frac{1}{2}\, m\, v^2\right) + \text{Potential energy}\,(m\,g\,h) = k\,.$$

We can evaluate this constant if we know the kinetic energy and the potential energy at any one height.

We know the potential energy at all heights and that the kinetic energy at $h_0 = 12$ m is zero, so

$$\frac{1}{2} m \, (0 \text{ m/s})^2 + m \, g \, h_0 \; = \; k.$$

Thus $k \; = \; m \, g \, h_0$ and at any height

$$\frac{1}{2} m \, v^2 + m \, g \, h \; = \; k \; = \; m \, g \, h_0,$$

giving

$$v^2 \; = \; 2 \times g \, (h_0 - h).$$

Then at $h \; = \; 10$ m.

$$v^2 \; = \; 2 \times 9.8 \text{ m/s}^2 \times (12 \text{ m} - 10 \text{ m})$$

$$v \; = \; \pm \, 6.26 \text{ m/s} \; = \; \pm \, 6.3 \text{ m/s}$$

considering significant figures. The $\pm$ is there since the ball is at 10 m both on its upward and its downward path. Note that the original height and the original velocity are not needed or even mentioned.

2. KINEMATICAL APPROACH

The kinematical equations of Chapter 2 of the text can be used to solve this problem. These equations were summarized in Table 2.1 of the text. There are two events discussed, the ball at 10 m and the ball at 12 m. Since we are not concerned with when the two events occurred, the last of the kinematical equations is most likely the one we want:

$$v^2 \; = \; v_0^2 + 2 \, a \, (x - x_0)$$

where v_0 is the velocity at x_0.

For our problem we replace x by y, x_0 by h_0, and a by $-g$. The minus sign is needed since y is measured positively upwards and the gravitational force is downwards. Then for $h_0 \; = \; 12$ m, $v_0 = 0$,

$$v^2 \; = \; -2 \, g \, (y - h_0) \; = \; 2 \, g \, (h_0 - y)$$

which is the same result as obtained with the energy conservation method. In this simple case there isn't much difference in effort, but often the energy method will be simpler. This is especially true in cases where the acceleration is not uniform or where rotational motion is involved (see Problem SG 6-5).

PROBLEM SG 6.7 An Atwood's machine has a mass of 10 kg hanging on the right side and a mass of 5.0 kg on the other. The pulley is a disk with a mass of 4.0 kg and a radius of 8.0 cm. Assume the rope moves over the pulley without slipping and that the objects are not moving initially. What is the motion of the 10 kg mass?

SOLUTION:

> **Procedure:** Choose force or energy method. We choose energy approach. Require the conservation of total energy.

Figure SG 6.7a shows the Atwood's machine for this problem. This problem will force us to revise our model of an ideal rope. We are told that the rope does not slip on the pulley, so the pulley itself must turn. If the pulley starts from rest and accelerates, there must be forces on the pulley. Since the rope is the only thing attached to the edge of the pulley, the rope must provide these forces. This is only possible if the tension in the right side part of the rope is different than the tension in the left hand side. Obviously, for a real rope, if we carefully analyzed the frictional forces we could understand this behavior. For our model of a rope we must allow the tension to change across those regions for which the rope is in contact with another surface across which it does not slip.

The best approach to problems is to first decide if a force or an energy approach is most appropriate. In this chapter ,we will consider the energy method. In Chapter 7, we will solve this problem again using forces. For reference the forces are shown in this chapter also.

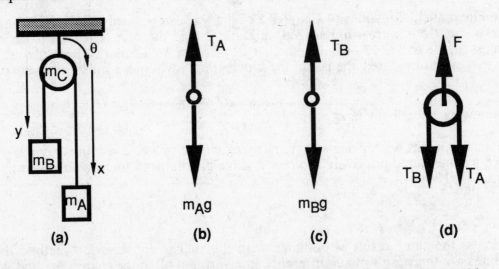

Figure SG 6.8:   a). The physical arrangement of the Atwood's machine.
b). Free body force diagram for body A.
c). Free body force diagram for body B.
d). The free body force diagram for the pulley.

We need to look first at the variables used and their interpretation.
The center of the pulley can serve as a reasonable reference point. We will let

x	indicate the distance of mass A from this point,
y	be the distance of mass B as shown in Figure SG 6.8a,
θ	be the angle through which the pulley has turned
R	will represent the radius of the pulley.

Since the rope has not stretched and does not slip on the pulley:

$$y - y_0 = -(x - x_0) \qquad \text{and} \qquad \theta - \theta_0 = \frac{x - x_0}{R}.$$

where y_0, x_0, and θ_0 represent the initial positions of the masses ans pulley. We will represent the time by t and take the initial time to be $t = 0$ and let

> v_y and a_y represent the velocity and acceleration of mass B,
> v_x and a_x be the velocity and acceleration of mass A, and
> ω and α the velocity and acceleration of the pulley.

The relations between y, θ and x give us

$$v_y = -v_x,$$

$$a_y = a_x,$$

$$\omega = \frac{v_x}{R},$$

and

$$\alpha = \frac{a_x}{R}.$$

These definitions and relations are given in Table SG 6.1

Table SG 6.1
Variables and Relations for Problem SG 6.7

object	position	velocity	acceleration	relation
mass A	x	v_x	a_x	
mass B	y	v_y	a_y	$v_y = -v_x$ $a_y = -a_x$
pulley	θ	ω	α	$\omega = v_x/R$ $\alpha = a_x/R$

GIVEN:

object	Symbol	Value
mass of body A	m_A	10.0 kg
mass of body B	m_B	5.0 kg
mass of pulley	m_C	4.0 kg
radius of pulley	R	8.0 cm
moment of inertia of pulley	I	$\frac{1}{2} m_C R^2$

NOT KNOWN:

position of mass A	x
velocity of mass A	v_x
acceleration of mass A	a_x
tension in rope on right	T_A
tension in rope on left	T_B
force on support tension in support rope)	F

ENERGY APPROACH:

We will take the zero of potential energy at the pulley. Then the

$$\text{initial potential energy} = -(m_A\, g\, x_0 + m_B\, g\, y_0)$$

and

$$\text{initial kinetic energy} = 0.$$

At any time the energy conservation theorem gives us:

$$-(m_A\, x_0 + m_B\, y_0)\, g = -(m_A\, x + m_B\, y)\, g + \frac{1}{2} m_A\, v_x^2 + \frac{1}{2} m_B\, v_y^2 + \frac{1}{2} I\, \omega^2 .$$

We also have

$$y - y_0 = -(x - x_0)$$

since as x increases y decreases by the same amount. Using this and the relations among v_x, v_y and ω, we get:

$$v_x^2 = 2 \times \frac{(m_A - m_B)\, g}{m_A + m_B + \dfrac{m_C}{2}} \times (x - x_0) = 2\, a_x\, (x - x_0),$$

where the last equality follows from Equation 2.7 of the text; that is, we can identify the coefficient of $(x - x_0)$ as the acceleration in the x-direction. If you enter the numbers for this problem, you will get the velocity of the 10 kg mass.

This result was obtained without reference to the forces on the masses or the pulleys.

PROBLEM SG 6.8: A 40 kg box is pushed at a constant velocity of 0.80 m/s with a horizontal force on a level floor. The coefficient of friction between the box and the floor is 0.500.

    (a). What is the applied force?
    (b). At what rate is power expended?
    (c). How much work is done in 1.0 minute?

SOLUTION:

> **Procedure:** Use forces. Since there is no vertical motion the vertical forces sum to zero and since the horizontal velocity is constant those forces also sum to zero.

The forces acting are gravity, the normal force and friction as shown in Figure SG 6.9. We use the following notation:

Quantity	Symbol	Value
Applied force	F	?
frictional force	F_f	?
mass of box	m	40 kg
acceleration of gravity	g	9.8 m/s^2
normal force	N	mg
velocity	v	0.80 m/s
coefficient of friction	μ	0.500

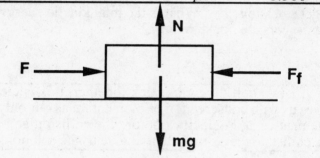

Figure SG 6.9:   Forces acting on box..

Since the box moves with constant horizontal velocity and has no vertical velocity:

$$N - mg = 0$$

$$F - F_f = 0.$$

We assume that the friction force is related to the normal force via the coefficient of friction, hence

$$F - F_f = F - \mu N = F - \mu\, mg = 0$$

so

$$F = \mu\, mg = 196\,\text{N} = 2.0 \times 10^2\,\text{N},$$

while

$$\text{Power} = \frac{\text{work}}{\text{time}} = \frac{\text{force} \times \text{distance}}{\text{time}}$$

$$= \text{force}\,\frac{\text{distance}}{\text{time}} = \text{force} \times \text{velocity}$$

$$= (2.0 \times 10^2\,\text{N}) \times 0.80\,\text{m/s} = 1.6 \times 10^2\,\text{W}.$$

while the work in one minute is

$$\text{Work} = \text{Power} \times \text{time} = 1.6 \times 10^2\,\text{N} \times 60\,\text{s}$$

$$= 9.6 \times 10^3\,\text{J}.$$

Or, if you prefer calculate the work in 1.0 minute by

$$\text{distance} = \text{velocity} \times \text{time} = 48 \text{ m}$$

then

$$\text{work} = 2.0 \times 10^2 \text{ N} \times 48 \text{ m} = 9.6 \times 10^3 \text{ J}$$

and

$$\text{power} = \frac{\text{work}}{\text{time}} = \frac{9.6 \times 10^3 \text{ J}}{60\text{s}} = 1.6 \times 10^2 \text{ W}.$$

PROBLEM SG 6.9: A 1200 kg automobile, going 100 m/s, collides head–on with a 1400 kg automobile which was traveling in the opposite direction at 50 m/s. During the collision the two cars get stuck together. What is the total kinetic energy immediately after the collision, assuming the cars are not spinning?

SOLUTION:

Procedure: Since the two colliding cars stick together, this is clearly not an elastic collision. That leaves us with conservation of momentum as our only conservation law. Since we know the masses and velocities before the collision and since there is only one object after the collision, that is sufficient to give us the solution.

We let

object	symbol	value
mass of car moving to right	m_1	1200 kg
velocity of car moving to right	v_1	100 m/s
mass of car moving to left	m_2	1400 kg
velocity of car moving to left	v_2	– 50 m/s
final velocity	v_f	?

A choice of positive direction has been made for car #1 so $v_2 < 0$; and, if $v_f > 0$ the final combination will be moving to the right in Figure SG 6.10b.

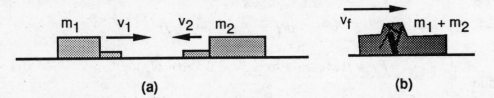

(a) (b)

Figure SG 6.10: a). The two cars before collision.
b). The two cars moving as a unit after collision.

Conservation of momentum is

$$\text{initial momentum} = \text{final momentum}$$

which is

$$m_1v_1 + m_2v_2 = (m_1 + m_2)v_f.$$

This gives

$$v_f = \frac{m_1v_1 + m_2v_2}{m_1 + m_2}$$

$$= 19 \text{ m/s}.$$

Since there is no rotation after the collision, the entire kinetic energy is translational and

$$\text{Final kinetic energy} = \tfrac{1}{2}(m_1 + m_2)v_f^2$$

$$= 4.8 \times 10^5 \text{ J}.$$

If there had been some final rotation, the value of v_f we calculated would have been correct and the final kinetic energy would have been correct for the translational energy, but there would also have been rotational kinetic energy.

The initial kinetic energy was about 6.8×10^6 J. The difference between this value and our calculated final value represents the work done to deform the cars and heat the materials.

PROBLEM SG 6.10: What velocity must a rocket have at the surface of the Earth if it is to rise at a height equal to the Earth's radius before it begains to desend? Use energy conservation.

SOLUTION:

Procedure: The conservation of energy will enable us to determine the height at which the kinetic energy is zero. Set this height equal to the Earth's radius.

If v is the initial velocity, m the mass of the rocket (assumed constant), m_E the mass of the Earth and R_E the radius of the Earth, then the initial total energy is

$$\tfrac{1}{2}mv^2 - G\frac{m\, m_E}{R_E}$$

and the total energy at $2R_E$, where the rocket stops before returning is

$$- G\frac{m\, m_E}{2R_E}$$

Conservation of energy makes these two equal. This gives

$$v = \sqrt{G\frac{m_E}{R_E}} = \sqrt{gR_E}\,,$$

where g is the acceleration of gravity at the Earth's surface.

PROBLEM SG 6.11: A box of mass m slides on a frictionless loop-the-loop starting from rest at a distance h_0 above the bottom. Let the loop radius be R. What is the minimum value that h_0 can have so that the box will not fall off the the loop?

SOLUTION:

Procedure: Since the normal force does no work, energy is conserved. Calculate the kinetic energy at the top of the loop and thus find the centrifugal force at that point. The centrifugal force must be larger than the gravitational force at that point.

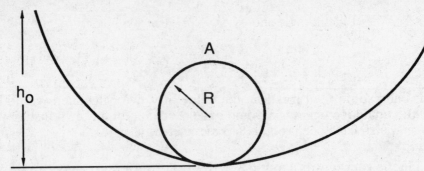

Figure SG 6.9: Loop-the-loop with a loop radius of R. A box enters the path from a height h_0.

We can use conservation of energy since the normal forces do no work. The total energy of the box is equal to its initial energy which is

$$mgh_0.$$

At the top of the loop, the box is moving horizontally to the left with a velocity v. The total energy at that point is

$$\frac{1}{2}mv^2 + mg(2R),$$

so

$$v^2 = 2g(h_0 - 2R).$$

The centifugal force needed for this motion is

$$\frac{mv^2}{R} = 2mg\left(\frac{h_0}{R} - 2\right).$$

The centrifugal force is provided by the normal force minus the gravitational force,

$$\text{centrifugal force} = \text{normal force} - mg,$$

so

$$\text{normal force} = \frac{2mgh_0}{R} - 5mg.$$

The condition that the box not leave the loop is that the normal force be greater than zero, that is

$$\frac{2mgh_0}{R} > 5mg$$

or

$$h_0 > \frac{5}{2}R.$$

Chapter 7
Linear and Angular Momentum

Important Terms

Fill in the blanks with the appropriate word or words.

1. A body behaves as if all its mass were concentrated at a point called the _____.

2. The _____ takes into account both the force acting and the placement of this force.

3. Momentum associated with translational motion is called _____ momentum.

4. A collision occurring between two bodies which stick together is called a(n) _____ collision.

5. The _____ of a body is the time rate of change of the body's angular velocity.

6. The total linear momentum before a collision is equal to the total linear momentum after the collision. This is a statement of the _____ .

7. A body who's angular motion is constant is said to be in _____ .

8. The _____ is the product of the linear momentum and the distance through which it acts.

9. The product of a force and the time that force acts is called a(n) _____ .

10. The quantity analogous to mass in angular motion is called _____ .

11. The total angular momentum before a collision is equal to the total angular momentum after the collision. This is a statement of the _____ .

Write the definitions of the following:

12. center of mass (7.8):

13. torque (7.7):

14. angular momentum (7.9):

15. conservation of angular momentum (7.10):

16. rotational equilibrium (7.8):

17. impulse (7.2):

18. angular acceleration (7.11):

19. conservation of linear momentum (7.3):

20. moment of inertia (7.9):

21. perfectly inelastic collision (7.4):

22. linear momentum (7.1):

Answers to 1-11, Important Terms

1. center of mass; 2. torque; 3. linear; 4. perfectly inelastic; 5. angular acceleration; 6. conservation of linear momentum; 7. rotational equilibrium; 8. angular momentum; 9. impulse; 10. moment of inertia; 11. conservation of angular momentum

General Comments

All of the text equations which involve acceleration are valid only when the acceleration is constant. Often we can model motion with changing acceleration provided the motion has relative long periods of constant acceleration and only relatively short periods when the acceleration changes. The model is to neglect the motion that occurs when the acceleration is changing.

In the preceding chapter we dealt with perfectly elastic collisions where the total kinetic and potential energy is conserved. In such collisions momentum is also conserved. This chapter introduces inelastic collisions which only conserve momentum. Momentum is always conserved while the total of kinetic and potential energy may not be conserved. We will later see that this missing energy shows up as heat or other forms of energy. With this broader definition of energy, we may also say that energy is always conserved; however, at this point in the course, you will only be able to use energy conservation when you know that the collisions are perfectly elastic.

Equation Analogies

The number of equations which you must remember can be reduced by considering the linear and the rotational equations as analogies. To help you remember these you should write the analogous angular equation for each of the given linear equations; however, you should first write the interpretation of the symbols used (fill in missing entries):

SYMBOLS USED FOR CONSTANT LINEAR OR ANGULAR ACCELERATION

symbol	interpretation	symbol	interpretation
x	linear position	θ	angular position
v	linear velocity	ω	angular velocity
a	linear acceleration	α	angular acceleration
F	force	τ	torque
m	mass	I	moment of inertia
p	linear momentum	L	angular momentum
KE_{trans}	linear kinetic energy	KE_{rot}	angular kinetic energy
Δ	indicates a change	Δ	indicates a change
t	time	t	time

The subscript o indicates the value of the quantity at $t = t_o$.

ANALOGOUS EQUATIONS

Linear	Rotational
$\bar{v} = \dfrac{\Delta x}{\Delta t}$	
$\bar{a} = \dfrac{\Delta v}{\Delta t}$	
$v^2 = v_0{}^2 + a(x - x_0)$	
$v = v_0 + a(t - t_0)$	
$x = x_0 + v_0(t - t_0)$	
$x = x_0 + v_0(t - t_0) + \dfrac{1}{2} a(t - t_0)^2$	
$\Sigma F_i = 0$	
$p = m v$	
$KE_{trans} = \dfrac{1}{2} m v^2$	

Use Table 7.1 of the text to check and correct your entries.

Procedures for Equilibrium Problems Involving Torque

The following general procedures will usually make equilibrium problems with torques reasonably easy to solve. These usually are problems in which extended bodies are supported by wires, pivots, brackets, and/or studs. A wire can provide a force only along its length while a pivot can provide a force in any direction but allows rotation. A bracket provides support and prevents rotation. A stud is assumed to be able to provide a force only perpendicularly to its surface.

1. Identify all forces acting. There will be gravity on the center of the object and forces at each support point.
2. Resolve all forces into horizontal and vertical components. Some of these components are known and others are not. Depending upon the questions asked you may need to find only some of the unknown force components. You will need at least as many equations among the force components as there are needed unknown components.
3. Translational equilibrium gives two equations relating the forces which you should always use:
   Sum of horizontal forces = 0
   Sum of vertical forces = 0.
4. Write as many additional torque equations as are needed. Choose the points about which you calculate torques carefully. They need to be chosen so as to be as simple as possible and if possible not to involve the force components that you neither know nor need to know. Usually the best points to use are those for which one or more force components have a zero length action arm. Such points have fewer terms in the torque equation.
5. If a careful choice of points for torques does not eliminate unnecessary unknowns, additional torque equations may be needed.
6. Solve the equations for the unknowns.

Sample Solutions

PROBLEM SG 7.1: A 0.14 kg baseball was thrown with an average speed of 30 m/s. The ball is hit with a bat and leaves the batting area with a speed of 35 m/s at an angle of 45° upwards with respect to the ground and in the reverse direction of its approach. What is the impulse given to the ball? If the duration of the contact between bat and ball is 20 ms, what is the average force? Assume the ball was traveling horizontally when hit.

SOLUTION:

> **Procedure:** Use law that change in momentum is equal to the applied impulse.

The simplest model will ignore all details of the ball except its mass which we will represent by m. The initial direction of the ball will be taken as +x and its final direction is 45° to the initial and with negative x and positive y components as shown in the motion diagram Figure SG 7.1a. We let v_0 be the initial velocity and v_1 be the velocity after impact with associated momenta p_0 and p_1. The relation we need is the definition of impulse:

$$\text{impulse} = F\,\Delta t = \Delta p.$$

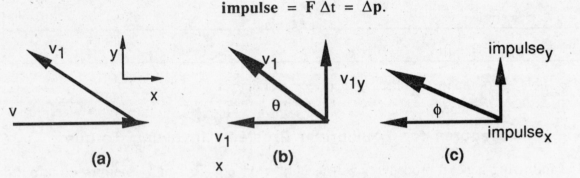

(a)   (b)   (c)

Figure SG 7.1:   a). Motion diagram for ball of problem SG 7.1.
b). Velocity components of final velocity.

GIVEN:

initial velocity	v_0	30 m/s
final velocity	v_1	35 m/s
final angle	θ	45°
duration of impulse	Δt	20 ms
mass of ball	m	0.14 kg

FIND:

impulse	**impulse**
average force	F

The component equations are

$$\text{impulse}_x = p_{x\ \text{initial}} - p_{x\ \text{final}}$$

$$= m\,v_0 - m(-v_1 \times \cos(\theta)) = m\,(v_0 + v_1 \times \cos(\theta))$$

$$= 0.14 \text{ kg} \times (30 \text{ m/s} + 0.707 \times 35 \text{ m/s}) = 7.6643 \text{ N s} = 7.7 \text{ N s}$$

$$\text{impulse}_y = 0 - m(-v_1 \times \sin(\theta)) = m\,v_1 \sin(\theta)$$

$$= 0.14 \text{ kg} \times 35 \text{ m/s} \times 0.707 = 3.4643 \text{ N-s} = 3.5 \text{ N-s}.$$

So the total impulse is 7.7 N-s in the $-x$ direction and 3.5 N-s in the $+y$ direction. These components are shown in Figure SG 7.1c.

The total impulse is thus

$$\text{impulse} = \sqrt{(7.7 \text{ N s})^2 + (3.5 \text{ N s})^2} = 8.5 \text{ N s}$$

at an angle of

$$\phi = \tan^{-1}(3.5/7.7) = 24°$$

to the $-x$-axis. The average force is

$$\frac{8.5 \text{ N s}}{20 \times 10^{-3} \text{s}} = 4.3 \times 10^2 \text{ N at } 24° \text{ up from } -x.$$

PROBLEM SG 7.2: A bird of mass 0.10 kg flying horizontally at 1.5 m/s is hit by a blob of tar of mass 0.050 kg falling vertically downward at 3 m/s. The blob sticks to the bird. What is the velocity of the bird immediately after the collision?

SOLUTION:

Procedure: An inelastic collision. We can not conserve energy but momentum is conserved. Only one body exists after the collision, so there is sufficient information in the one conservation law.

We will choose $+x$ to be the initial direction of the bird and $+y$ to be vertically downward. Since all information is given just before and just after the collision, we can neglect the effects of gravity, air resistance and the bird's wing movement. The three velocities are shown in Figure SG 7.2a while the components and direction of the final velocity is shown in Figure SG 7.2b.

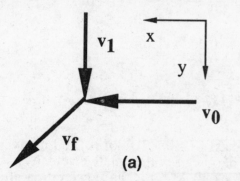

(a)

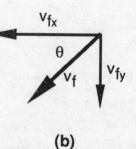

(b)

Figure SG 7.2:   a) Initial velocities and final velocity.
                       b) components of final velocity.

GIVEN:

Quantity	symbol	value
initial velocity of bird	v_0	1.5 m/s in $+x$ direction
initial velocity of blob	v_1	3.0 m/s in $+y$ direction
mass of bird	M	0.10 kg
mass of tar blob	m	0.050 kg

FIND:

$$\text{final velocity} \qquad v_f$$
$$\text{final direction} \qquad \theta$$

Conservation of momentum gives us:

$$\text{Initial momentum} = \text{final momentum}$$

$$M\, v_0 + m\, v_1 = (M + m)\, v_f\,.$$

In terms of components:

$$M\, v_0 = (m + M)\, v_f \times \cos(\theta),$$

$$m\, v_1 = (m + M)\, v_f \times \sin(\theta).$$

Using

$$\sin^2(\theta) + \cos^2(\theta) = 1$$

and

$$\tan(\theta) = \frac{\sin(\theta)}{\cos(\theta)}$$

we get:

$$v_f = \frac{\sqrt{M^2 v_0^2 + m^2 v_1^2}}{m + M} = \frac{0.212 \text{ kg m/s}}{0.15 \text{ kg}} = 1.4 \text{ m/s}$$

and

$$\theta = \tan^{-1}\left(\frac{m\, v_1}{M\, v_0}\right) = \tan^{-1}\left(\frac{0.15}{0.15}\right) = 45°.$$

The bird is now moving forward and downward at an angle of 45°.

PROBLEM SG 7.3: A rocket with an initial mass of 1.0×10^4 kg is fired in a vertical direction. Its exhaust gases are ejected at the rate of 50 kg/s with a relative velocity of 3.0×10^3 m/s. What is the initial acceleration of the rocket? What is the acceleration after 30 seconds have passed? What is the acceleration after 2 minutes have passed? What reasons do you expect exist that makes us avoid asking for the velocity or the position at 30 s and 120 s? Assume that the rocket was initially at rest on the Earth's surface and that the rocket does not run out of fuel during the time considered.

SOLUTION:

Procedure: The first parts of this problem are a straightforward application of the thrust relation of the text. Identify the forces acting on the rocket and use Newton's Second Law.

Figure SG 7.3: Rocket with thrust upwards and gravity pulling downwards.

The solution is similar to example 7.7 of the text. We model the rocket as a point and consider the forces acting on the rocket. These are gravity acting downward and the rocket thrust acting upwards as shown in Figure SG 7.3. We let F_R represent the rocket thrust and F_G be the gravitational force. The thrust is given by Equation 7.9 of the text as

$$F_R = v_r \frac{\Delta m}{\Delta t}$$

where

$$\frac{\Delta m}{\Delta t}$$

is the rate that the rocket ejects gases and v_r is the ejection velocity relative to the rocket. The gravitational force will be $F_G = -m g$ where m is the mass of the rocket plus remaining fuel and g is the acceleration due to gravity. The minus sign shows that F_R and F_G are in opposite directions. We let m_0 be the initial mass of rocket plus fuel.

GIVEN:

Quantity	symbol	value
ejection rate	$\dfrac{\Delta m}{\Delta t}$	50 kg/s
ejection speed	v_r	3.0×10^3 m/s
initial mass	m_0	1.0×10^4 kg

If we let $t = 0$ be the time of launch, then the mass of rocket plus remaining fuel at any time t, $m(t)$, will be given by:

$$m(t) = m_0 - \frac{\Delta m}{\Delta t} t.$$

The physical law we need is Newton's third law that the net force is equal to the mass times the acceleration. We let a represent the acceleration in the x direction, taken positively upwards and can now write:

$$F_{total} = F_R + F_G = v_r \frac{\Delta m}{\Delta t} - m(t) g = m(t) a.$$

This gives

$$a = \frac{v_r}{m(t)} \frac{\Delta m}{\Delta t} - g$$

$$= \frac{v_r}{m_0 - \frac{\Delta m}{\Delta t} t} \frac{\Delta m}{\Delta t} - g.$$

For our problem this is

$$a = (\frac{15}{1 - 5.0 \times 10^{-3} \times t} - 9.8) \text{ m/s}^2,$$

where t is in seconds. Hence

$$a(0 \text{ s}) = 5.2 \text{ m/s}^2,$$

$$a(30 \text{ s}) = 7.8 \text{ m/s}^2,$$

and

$$a(120 \text{ s}) = 28 \text{ m/s}^2.$$

Since the acceleration is not constant we cannot use the relations in Chapter 2 to calculate the velocity and the displacement. Calculus is needed to exactly solve problems with varying acceleration. You have the background to get numerical answers to the position and velocity for problems such as this. The following steps would be needed:

1. Plot the acceleration as a function of time.

2. For any time t, the total area under the acceleration curve, up to time t, is the velocity at that time t.

3. Plot the calculated velocities against time. The displacement is the total area under the velocity curve up to the time t.

PROBLEM SG 7.4: A square sign one meter on a side and of mass 50 kg is suspended from a wall by a bracket at the upper corner, a stud at the lower corner, and a support wire as shown in Figure SG 7.4a. The wire makes an angle of 30° with respect to the horizontal at the outer edge of the sign and has a tension of 40 N. What is the force provided by the upper bracket and how much force does the stud exert on the sign?

SOLUTION:

> **Procedure:** Follow the procedures for equilibrium problems involving torque. The first two steps are to identify the forces and resolve them into components.

We label the tie point for the wire by the letter A and the upper and lower supports by B and C respectively. The length of each side of the sign is denoted by the symbol d. The center of mass of the sign will be taken to be in the geometrical center of a perfect square one meter on a side. The bracket will be assumed to be able to provide any force needed. Details of the bracket will be ignored.

In Figure SG 7.4b we show the free body diagram for the sign. The bracket can provide an upward or downward force and a horizontal force. Rather than putting in an arbitrary force and angle, it is usually easier to put in the vertical and horizontal components and calculate the force from these at the last step or just give these components as the answer. The positive direction of the forces has been arbitrarily chosen as shown. If the magnitude of any force comes out negative, then the force points in the opposite direction to the that shown.

The unknown forces we want are V_B, F_B, and F_C, thus we need 3 equations among the force components.

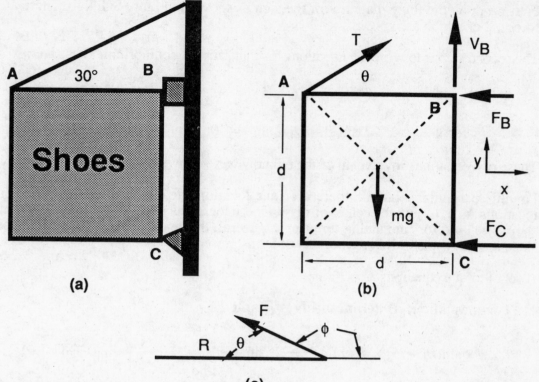

Figure SG 7.4: a). The physical arrangement of sign, wall, brackets, and tension wire.

b). A free body force diagram for the sign. The dotted lines are the diagonals of the sign and the bold arrows are the forces. Each force is shown acting at its proper location.

c). Torque produced about C by force F may be calculated as either $F \times R \times \sin \theta$ or as $F \times R \times \sin \phi$.

The sign should be in both translational and rotational equilibrium. If we let F_i denote any of the forces and τ_i be the torques calculated about any point then these equilibria conditions are:

$$\sum_i F_i = 0 \qquad \text{and} \qquad \sum_i \tau_i = 0.$$

Summing the forces will present no problems; however, to sum the torques, you must decide which and how many centers to use. The following rules provide a good guide:

1. Determine how many unknown forces need to be determined. (For this problem there are 3 forces that are not known, V_B, F_B, and F_C, since m and T are given.)
2. Since translational equilibrium will give us two equations (considering two dimensional problems), subtract two from the number of unknown forces. (Leaving 1 for this problem)
3. The remaining number is the number of torque equations needed. (for this problem there is 1).
4. Choose the centers needed so as to produce the simplest equations to solve. This is the hardest part to do without experience. Usually the best choices are points at which forces act, since these choices will eliminate one or more forces from the torque equation(s); however, any choice(s) will work. (For this problem we could choose points A or B or C.)

First we draw the force diagram for the sign as shown in Figure SG 7.4b. There are 5 forces, 2 of which are known.

The second step is to write the equations for translational equilibrium. These are:

$$\text{for } x \qquad T\cos\theta - F_B - F_C = 0 \qquad\qquad \text{(SG 7.1)}$$

$$\text{for } y \qquad T\sin\theta + V_B - mg = 0. \qquad\qquad \text{(SG 7.2)}$$

These two equations involve all of the unknowns, so we will need at 1 torque equation.

For this particular example there are four locations at which forces act. They are locations A, B, C, and the center of mass. These locations are among the centers which should be used for calculating torques. In general, the center of mass is a poor choice since it usually maximizes the number of terms present.

A set of torque equations are

Torques about B (eliminating V_B and F_B).

$$T \times d \times \sin(\theta) + F_C \times d - m\,g\,\frac{\sqrt{2}\,d}{2} \times \sin(135°) = 0 \qquad\qquad \text{(SG 7.3)}$$

where torques are taken as positive if clockwise (arbitrary choice).

At this point some additional comments on writing torques are in order. First, since $\sin(\Omega) = \sin(90° + \Omega)$ we could have written Equation SG 7.3 as

$$T \times d \times \sin(\theta) + F_C \times d - m\,g\,\frac{\sqrt{2}\,d}{2} \times \sin(45°) = 0. \qquad\qquad \text{(SG 7.3')}$$

This choice is shown in Figure SG 7.4c. Secondly, we could have also projected the gravitational force on the dotted line and immediately written Equation SG 7.3 as

$$T \times d \times \sin(\theta) + F_C \times d - m\,g\,\frac{d}{2} = 0. \qquad\qquad \text{(SG 7.3'')}$$

All of these are equivalent, the latter obviously is simplest.
Another possible choice is

Torques about A (eliminating T and F_B)

$$-V_B \times d + F_C \times d = 0. \qquad\qquad \text{(SG 7.4)}$$

Equation SG 7.4 gives
$$F_C = V_B \qquad\qquad \text{(SG 7.5)}$$
while SG 7.3 gives
$$F_C = \frac{mg}{2} - T\sin\theta. \qquad\qquad \text{(SG 7.6)}$$

Equations SG 7.1 and 7.2 give, respectively

$$T\cos\theta = F_B + F_C \qquad\qquad \text{(SG 7.7)}$$

and

$$T \sin \theta = mg - V_B.$$ (SG 7.8)

The simplest choice is to use SG 7.8 to find V_B, then SG 7.5 gives F_C and SG 7.7 gives F_B. That is,

$$V_B = mg - T \sin \theta$$

$$F_C = V_B$$

and

$$F_B = T \cos \theta - F_C.$$

From the numbers given for this problem, we find

$$V_B = 4.7 \times 10^2 \text{ N}$$
$$F_C = 4.7 \times 10^2 \text{ N}$$
$$F_B = -4.4 \times 10^2 \text{ N}.$$

The force F_B is in the opposite direction from that shown in Figure SG 7.4b.

PROBLEM SG 7.5: A plastic "flying bird" consists of a plastic body on a string. The string passes through a hollow tube (like a soda straw) and is held at the other end. The tube is held vertically and moved rapidly in a small horizontal circle to make the "bird" fly as shown in Figure SG 7.5. The "bird" has a mass of 80 g. The "bird" is set "flying" in a 50 cm radius and the straw is held fixed. If the string is pulled down in the straw so that the radius is now 15 cm, what is the fractional change in the period of the "bird" about the straw? Assume the "bird" is moving slowly enough that we can neglect the lift provided by the air. This assumption makes for poor flying, but it gives us a simpler problem to solve.

SOLUTION:

> **Procedure:** Since the string provides only a radial force the angular momentum is conserved when the string is shorted, so we will use conservation of angular momentum. Since we are asked to find only the fractional change, it is not necessary to solve for the frequency.

We will consider the "bird" to be a small mass of mass m and the string to be ideal. The motion of the straw is just to get the bird flying so we will consider the straw as fixed while the string length is changed. The physical arrangement is shown in Figure SG 7.5a.

We can answer the question without finding the frequency of motion. For the "bird" the initial angular momentum, L_i, was

$$L_i = m R_i^2 \omega_i$$

while the final angular momentum, L_f, is

$$L_f = m R_f^2 \omega_f,$$

where R and ω are the radii and frequency respectfully.

Since the string exerts only a radial force, the initial and final angular momentum are equal, giving:

$$m R_i^2 \, \omega_i = m R_f^2 \, \omega_f$$

or

$$\frac{\omega_f}{\omega_i} = \frac{R_i^2}{R_f^2} = \frac{(50 \text{ cm})^2}{(15 \text{ cm})^2} = 11.$$

The final angular frequency is 11 times the initial angular frequency.

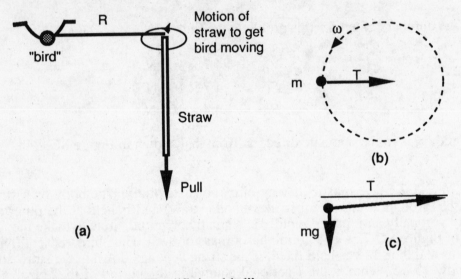

Figure SG 7.5: a). The "flying bird" on a straw.
b). Forces in the horizontal plane.
c). Forces in a vertical plane.

PROBLEM SG 7.6: Suppose in the preceding problem the The string makes an angle of 10° below the horizontal, what is the force that is exerted to the string to keep the "bird" at 15 cm radius and what is the frequency at which the "bird" revolves? Neglect lift provided by the air.

SOLUTION:

Procedure: Now we must use the equality of the horizontal tension of the string to the centripetal force and the vertical tension to the gravitational force.

We can answer this question with material from Chapter 3.
Equation 5.6 relates the centripetal force, F_C, and the resulting circular motion:

$$F_C = m \frac{v^2}{r} = m \, \omega^2 r \qquad \text{(centripetal force)}$$

where r is the radius of the circle, v is the tangential velocity, and ω is the angular frequency

As we see from Figure SG 7.5c, letting θ be the angle from the horizontal,

$$m \, g = T \sin(\theta) \qquad \text{(vertical forces)}$$

and from Figure SG 7.5b,

$$F_C = T \cos(\theta) = m \, \omega^2 r. \qquad \text{(horizontal forces)}$$

Hence,

$$T = \frac{m \, g}{\sin(\theta)} = \frac{0.080 \times 9.8}{0.178} \, N = 44 \, N \, ,$$

while

$$\omega = \sqrt{\frac{T \cos(\theta)}{m \, r}} = \sqrt{\frac{g \cos(\theta)}{r \sin(\theta)}}$$

$$= \sqrt{\frac{9.8 \times 0.9848}{0.15 \times 0.1736}} \, \text{rad/s} = 19 \, \text{radians/s.}$$

PROBLEM SG 7.7:   A wheel consists of a hub of mass 0.10 kg and diameter 10 cm, a set of 8 spokes and an outer rim 1.2 m in diameter. The outer rim is of iron and has a mass of 50 kg. The wheel is initially spinning on an axis bout the hub at a rate of 100 revolutions per minute without friction. If a force of 10 N is applied tangentially to the wheel at the rim, how long will it be before the wheel stops and through how many revolutions will it turn? Neglect the mass of the spokes.

SOLUTION:

Procedure: This is an application of the equations relating angular motion and angular acceleration. First calculate the moment of inertial of the wheel. There is torque due to the applied force. This torque causes an angular acceleration. The angular acceleration stops the wheel.

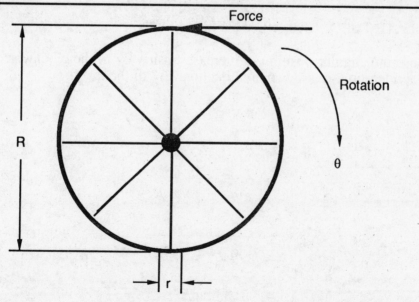

Figure SG 7.6:   Wheel of radius 0.60 m with constant stopping force. The angular position, θ is measured positively in a clockwise sense as shown.

The problem is illustrated in Figure SG 7.6. We will assume the hub is a uniform cylinder of radius $r = 5.0$ cm and mass, $m = 0.10$ kg while the rest of the wheel is a ring of radius $R = 0.60$ m and mass $M = 50$ kg.

We will represent the initial rotation of the wheel by $\omega_0 = 2\pi \times$ frequency $= 2\pi \times$ 100 rev/min $\times$ 1min/60 s $= 10.5$ rad/s. The applied torque is constant and will produce a constant angular acceleration, α, which will reduce the rotation rate of the wheel. Hence we are

GIVEN:

Quantity	symbol	value
radius of hub	r	0.05 m
mass of hub	m	0.10 kg
radius of rim	R	0.6 m
mass of rim	M	50 kg
applied force	F	10 N
applied torque	$\tau = -FR$	-6 N m
initial rotation rate	ω_0	10.5 rad/s

CALCULATE THE MOMENT OF INERTIA, I_T, FOR THE WHEEL

The net moment of inertia of the system is

$$I_T = I_{hub} + I_{rim},$$

where

$$I_{hub} = \frac{1}{2} m r^2$$

and

$$I_{rim} = M R^2$$

as can be found in Table 7.3 of the text.

CALCULATE THE ANGULAR ACCELERATION

We choose the angular position to increase positively in the clockwise sense, so the angular acceleration, α, as given by Equation 7.17 of the text is

$$\alpha = \frac{\tau}{I_T}$$

$$= \frac{\tau}{\frac{1}{2} m r^2 + M R^2}$$

$$= \frac{R F}{\frac{1}{2} m r^2 + M R^2}$$

$$= -0.33 \text{ rad/s}^2,$$

where τ is the torque.

The effect of the hub is too small to consider with our significant figures.

CALCULATE THE ROTATION ANGLE

The rotational equation form Table 7.1 of the text is

$$\omega = \omega_0 + \alpha\, t.$$

The wheel will stop at a time given by

$$t = -\frac{\omega_0}{\alpha}$$

$$= -\frac{10.5\ \text{rad/s}}{-0.33\ \text{rad/s}^2} = 32\ \text{s}.$$

The wheel will stop in 32 s.

ANGLE OF TURN

This is most easily calculated by using the relation

$$\omega^2 = \omega_0^2 + 2\,\alpha\,(\theta - \theta_0),$$

We know $\omega = 0$ and $\omega_0 = 10.5$ /s., and where θ is the angle of rotation This gives

$$\Delta\theta = \theta - \theta_0 = \frac{\omega^2 - \omega_0^2}{2\alpha} = -\frac{\omega_0^2}{\alpha} = 167\ \text{rad}.$$

Each revolution is 2π radians so this is about 27 revolutions.

PROBLEM SG 7.8: An Atwood's machine has a mass on 10 kg hanging on the right side and a mass of 5.0 kg on the other. The pulley is a disk with a mass of 4.0 kg and a radius of 8.0 cm. Assume the rope moves over the pulley without slipping and that the objects are not moving initially. What is the motion of the 10 kg mass?

SOLUTION:

> **Procedure:** This problem was solved using energy considerations in the last chapter. You now have the material to solve this by force considerations. It is presented this way to force you to consider how ropes behave when friction is present.

This is a repeat of problem SG 6.7, which will be solved in a different fashion in this chapter. Figure SG 7.7a shows the Atwood's machine for this problem. This problem forces us to revise our model of the ideal rope. We are told that the rope does not slip on the pulley, so the pulley itself must turn. If the pulley starts from rest and accelerates, there must be a torque on the pulley. Since the rope is the only thing attached to the edge of the pulley, the rope must provide this torque. This is only possible if the tension in the right side part of the rope is different than the tension in the left hand side. Obviously, for a real rope, if we carefully analyzed the frictional forces we could understand this behavior. For our model of a rope we must allow the tension to change when friction is considered.

The best approach to problems is to first decide if a force or an energy approach is most appropriate. This problem is most simply solved by energy considerations; however, it is instructive to consider a force approach. Consider the variables used and their interpretation. The center of the pulley can serve as a reasonable reference point. We will let x indicate the distance of mass A from this point and y be the distance of mass B as shown in Figure SG 7.8a. We will also let θ be the angle through which the pulley has turned. R will represent the radius of the pulley.

Since the rope has not stretched and does not slip on the pulley:

$$y - y_0 = -(x - x_0)$$

and

$$\theta - \theta_0 = \frac{x - x_0}{R}.$$

where y_0, x_0, and θ_0 represent the initial positions of the masses and pulley. We will represent the time by t and take the initial time to be $t = 0$. We will also let v_y and a_y represent the velocity and acceleration of mass B, v_x and a_x be the velocity and acceleration of mass A, and ω and α the velocity and acceleration of the pulley. The relations between y, θ and x give us:

$$v_y = -v_x, \qquad\qquad a_y = a_x,$$

and

$$\omega = v_x/R, \qquad\qquad \alpha = a_x/R.$$

These definitions and relations are given in Table SG 7.1

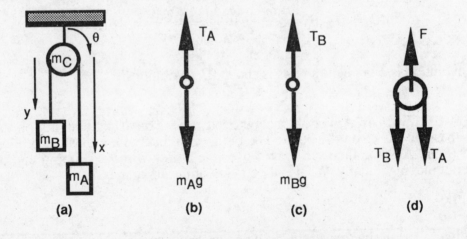

(a)   (b)   (c)   (d)

Figure SG 7.7:   a). The physical arrangement of the Atwood's machine.
b). Free body force diagram for body A.
c). Free body force diagram for body B.
d). The free body force diagram for the pulley.

Table SG 7.1
Variables and Relations for Problem SG 7.7

object	position	velocity	acceleration	relation
mass A	x	v_x	a_x	
mass B	y	v_y	a_y	$v_y = -v_x$ $a_y = -a_x$
pulley	θ	ω	α	$\omega = v_x/R$ $\alpha = a_x/R$

GIVEN:

quantity	symbol	value
mass of body A	m_A	10 kg
mass of body B	m_B	5.0 kg
mass of pulley	m_C	4.0 kg
radius of pulley	R	8.0 cm

UNKNOWN:

Quantity	Symbol
position of mass A	x
velocity of mass A	v_x
acceleration of mass A	a_x
tension in rope on right	T_A
tension in rope on left	T_B
force on support (tension in support rope)	F

Figures SG 7.7b–d show the forces and torques acting on the three bodies of the problem. We let F be the force on the string holding the pulley up, then from Newton's Second Law for translational and rotational motion:

Translational:

$$m_A g - T_A = m_A a_x$$

$$m_B g - T_B = m_B a_y = -m_B a_x$$

$$F - T_B - T_A = 0$$

Rotational:

$$(T_A - T_B) \times R = I \alpha = I \frac{a_x}{R} = \frac{1}{2} m_C R^2 \frac{a_x}{R}$$

$$= \frac{m_C \times R \times a_x}{2},$$

where we have used the moment of inertia, I, for a disk about its axis as given in Table 7.3 of the text. The first, second and last equations together give us:

$$a_x = \frac{m_A - m_B}{m_A + m_B + \frac{m_C}{2}} g = 2.9 \text{ m/s}^2.$$

Note that these results are independent of R.

Since the initial velocity was zero and the initial time was taken as zero,

$$x(t) = x_0 + \frac{1}{2} a_x t^2 = x_0 + 1.5 \text{ m/s}^2 \times t^2.$$

This allows us to find the position at any time.

PROBLEM SG 7.9: An Atwood's machine has a mass of 12 kg hanging on one side and a mass of 7.0 kg on the other side. The pulley is a disk with a mass of 7.0 kg and a radius of 8.00 cm. Assume the string connecting the two masses does **not** slip on the pulley. (a). If the masses start from rest, how fast will the 12 kg mass be moving when it has fallen 80 cm? (b). How long will this take?

SOLUTION:

> **Procedure:** Use energy considerations to find the velocity after the masses have moved 80 cm. From this, and the basic kinematical laws, the acceleration and time can be determined.

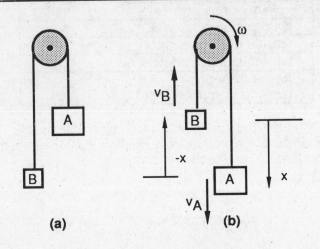

(a)   (b)

Figure SG 7.8:   a) Atwood's machine when masses released.
                        b) The system after a time t.

The initial state is shown in Figure SG 7.8a and the system after it has fallen 80 cm is shown in Figure SG 7.8b. We denote the two masses by A and B. Since they are connected by a string, the amount A moves down is equal to the amount B moves up. At any point in the motion the energy is conserved; that is, if PE is gravitational potential energy and KE is kinetic energy:

$$PE_A + KE_A + PE_B + PE_B + KE_{pulley} = constant.$$

Since the system starts from rest the initial kinetic energy is zero and if we measure the positions of the two masses by x, the displacement from the starting position, then we may choose this constant to be zero.

$$PE_A + KE_A + PE_B + PE_B + KE_{pulley} = 0.$$

In terms of x and the instantaneous angular velocity of the pulley this becomes

$$-m_A g x + \tfrac{1}{2} m_A v_A^2 + m_B g x + \tfrac{1}{2} m_B v_B^2 + \tfrac{1}{2} I \omega^2 = 0$$

where v_A and v_B are the velocities of mass A and mass B respectively and I is the moment of inertia of the pulley. The first and third terms are of opposite sign since if A and B move in opposite directions.

Since the string is assumed not to slip on the pulley,

$$\omega r = v_A = v_B$$

where r is the radius of the pulley, ω is the rotational velocity of the pulley, and the last equality follows from the fixed length of the string. Thus the energy equation becomes

$$-(m_A - m_B)\,gx + \frac{1}{2}(m_A + m_B)v_A^2 + \frac{1}{2}I\left(\frac{v_A}{r}\right)^2 = 0$$

or

$$v_A^2 = \frac{2(m_A - m_B)gx}{m_A + m_B + \dfrac{I}{r^2}}$$

$$= \frac{2(m_A - m_B)gx}{m_A + m_B + \frac{1}{2}m_{\text{pulley}}}$$

for $x = 0.80$ m, this gives

$$v_A = 1.9 \text{ m/s}$$

and since the mass started from rest, we can use $v^2 = 2ax$ to find the acceleration

$$a = \frac{v^2}{2} = 2.3 \text{ m/s}^2$$

and; hence, since $x = \frac{1}{2}a\,t^2$,

$$t = \sqrt{\frac{2x}{a}} = 0.83 \text{ s}$$

PROBLEM 7.10: A sign weighing 980 N is suspended from the end of a uniform rod 5.0 m long and weighing 1470 N. There is a support cable on the outer end which makes an angle of 45° with respect to the rod (as shown in Figure SG 7.9a)). What is the tension in the support cable?

SOLUTION:

Procedure: There is one unknown so one equation should solve this problem. If we take torques about the wall support for the wall, we can determine the tension.

For the rod to be in equilibrium the translational forces and rotational torques must each add to zero. There are four forces on the rod. Its own weight, the weight of the sign, the tension in the support cable, and the force provided by the wall-rod support. This is shown in Figure SG 7.9b. We are asked to find the tension in the wire. This is one unknown, which suggests that possibly only one equation is needed.

Since we are not asked for the rod-wall support force, we can get this one equation by taking torques about the rod-wall point. This equation, which will not involve the forces at that point, is

$$\text{(weight of rod)} \times \frac{L}{2} + \text{(weight of sign)} \times L - T \sin \theta \times L = 0,$$

where L is the length of the rod, θ is the angle between the rod and the cable, and we have assumed the rod acts as if all its mass is concentrated at the center. Thus,

$$T = \frac{\dfrac{\text{weight of rod}}{2} + \text{weight of sign}}{\sin \theta} = 2.4 \times 10^3 \text{ N}.$$

The cable has a tension of 2.4×10^3 N.

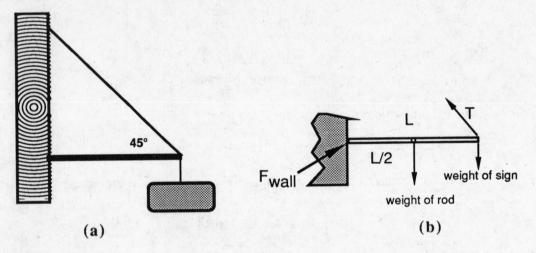

(a)	**(b)**

Figure SG 7.9:   a) A sign hanging on a rod with a 45° tension wire.
b) Forces on the rod.

PROBLEM SG 7.11: An electric motor is used to lift a 20-kg bucket of water at constant speed from a well . If the diameter of the pulley is 10.0 cm, what must be the torque of the motor?

SOLUTION:

> **Procedure:** At constant speed the force in the support rope must be the weight of the bucket. Calculate the torque for this force.

To lift the 20-kg mass at constant speed, we need a force equal and opposite to the force of gravity. This is a force

$$F = mg = 20 \text{ kg} \times 9.80 \text{ m/s}^2,$$

$$= 196 \text{ N}.$$

Equation (7.9) applies here with $\theta = 90°$, so

$$\tau = r F \sin 90° = rF,$$
$$= 0.050 \text{ m} \times 196 \text{ N} = 9.8 \text{ N–m}.$$

The motor must be able to provide a torque of 9.8 N-m

Chapter 8
Conservation of Energy and Momentum

Important Terms

Fill in the blanks with the appropriate word or words. Since this chapter consists of applications of the material in the last two chapters, Some important terms from those chapters are included here again.

1. The _____ is the unit of work in the SI system.

2. The time rate of doing work is called _____.

3. _____ is the energy of translational motion.

4. The ability to do work is called _____.

5. The energy that a body has because it is in a gravitational field is called the body's _____ energy.

6. The quantity which plays the same role as mass in the expression for rotational kinetic energy is called the _____.

7. _____ is defined as the product of a force times the distance through which the force acts.

8. The _____ states that the work done on a body is equal to its change in kinetic energy.

9. The _____ is a measure of the energy of rotational motion.

10. The ideal spring produces a force proportional to its stretching. This is called _____.

11. The SI unit of power is called the _____.

12. The sum of kinetic and potential energy is called _____.

13. Our first conservation law is the _____.

14. The _____ takes into account both the force acting and the placement of this force.

15. The product of a force and the velocity with which it is applied is the _____.

16. A collision which conserves energy is said to be a(n) _____ collision.

17. A body behaves as if all its mass were concentrated at a point called _____.

18. The _____ of a rocket is the force produced by ejecting gases.

19. Momentum associated with translational motion is called _____ momentum.

20. A collision occurring between two bodies which stick together is called a _____ collision.

21. A collision which does not conserve kinetic energy is called a(n) _____ collision.

22. The _____ of a body is the time rate of change of the body's angular velocity.

23. The total linear momentum before a collision is equal to the total linear momentum after the collision. This is a statement of the _____ .

24. A body who's angular motion is constant is said to be in _____ .

25. A _____ consists of a pair of forces not lying along the same line of equal magnitude but opposite direction.

26. The total velocity corresponding to the initial upward motion necessary for a body not too fall back to earth is called the _____ velocity

27. The quantity analogous to mass in angular motion is called _____ .

28. The _____ is the product of the linear momentum and the distance through which it acts.

29. The total angular momentum before a collision is equal to the total angular momentum after the collision. This is a statement of the _____ .

Write the definitions of the following:

30. work (6.1):

31. energy (6.2):

32. joule (6.1):

33. Hooke's Law (6.1):

34. gravitational potential energy (6.4):

35. inelastic collision (8.1):

36. work-energy theorem (6.1):

37. total mechanical energy (6.7):

38. perfectly elastic collision (8.1):

39. instantaneous power (6.8):

40. translational kinetic energy (6.8):

41. perfectly inelastic collision (8.1,7.4):

42. rotational kinetic energy (6.5):

43. power (6.8):

44. moment of inertia (6.6):

45. distinguish between work and power

46. black hole (8.7):

47. conservation of mechanical energy (6.7):

48. angular acceleration (7.11):

49. conservation of linear momentum (7.4):

50. moment of inertia (7.9):

51. escape velocity (8.7):

52. linear momentum (7.4):

53. watt (and its definition in terms of joule/s) (6.8):

54. center of mass (7.8):

55. torque (7.7):

56. couple (7.8):

67. angular momentum (7.9):

58. conservation of angular momentum (7.10):

59. rotational equilibrium (7.8):

60. thrust (7.6):

Answers to 1-29, Important Terms

1. joule; 2. power; 3. translational kinetic energy; 4. energy; 5. gravitational potential; 6. moment of inertia; 7. work; 8. work-energy theorem; 9. rotational kinetic energy; 10. Hooke's Law; 11. watt; 2. total mechanical energy; 13. conservation of mechanical energy; 14. torque; 15. instantaneous power; 16. elastic; 17. center of mass; 18. thrust; 19. linear; 20. perfectly inelastic; 21. inelastic; 22. angular acceleration; 23. conservation of linear momentum; 24. rotational equilibrium; 25. couple; 26. escape; 27. moment of inertia; 28. angular momentum; 29. conservation of angular momentum

Sample Solutions

PROBLEM SG 8.1: When a glass marble bounces off a steel plate 80.0% of the kinetic energy is conserved. If the marble is dropped from an initial height of 3.000 m, how high will it bounce after the second impact?

SOLUTION:

> **Procedure:** The momentum of the marble is not conserved upon collision, but upon each collision 80% of the kinetic energy is retained. Total energy is conserved between collisions. This means that at the top of each bounce the potential energy is 80% of what it was at the top of the last bounce.

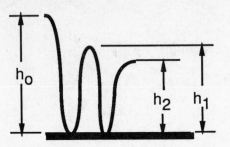

Figure SG 8.1:   Marble bouncing off a steel plate.

A motion plot is shown in Figure SG 8.1. We let h_0 be the initial height, h_1 be the height of the first bounce, and h_2 be the maximum height after the second bounce. Potential energy is measured from the plate. Since the system is interacting with the earth, we **can not** use conservation of momentum. Energy will be conserved except during the collisions when 20.0% of the energy is lost. We label events as follows:

EVENT LABEL	EVENT	ENERGY and TYPE
A	start at h_0	mgh_0 — potential
B	just before first impact	mgh_0 — kinetic
C	just after first impact	$0.8\, mgh_0$ — kinetic
D	at h_1	$mgh_1 = 0.8\, mgh_0$ — potential
E	just before second impact	mgh_1 — kinetic
F	just after second impact	$0.8\, mgh_1$ — kinetic
G	at h_2	$mgh_2 = 0.8\, mgh_1 = 0.64\, mgh_0$

From this analysis, we see that the maximum height after two bounces is just 0.64 of the initial height or 1.920 m. Notice that three significant figures were given.

A shorter analysis is to state that 80% of the initial energy remains after each collision, so that after two collisions, only 64% = 80% × 80% of the initial energy remains.

PROBLEM SG 8.2: An air track glider of 100 g mass with an initial speed of 6.0 m/s collides head on with another glider at rest. The other glider is 5.0 times as massive as the first one. What are the final speeds and directions of the gliders if the collision is 90% elastic?

SOLUTION:

Procedure: Assume conservation of momentum. Also after the collision only 90% of the initial kinetic energy remains.

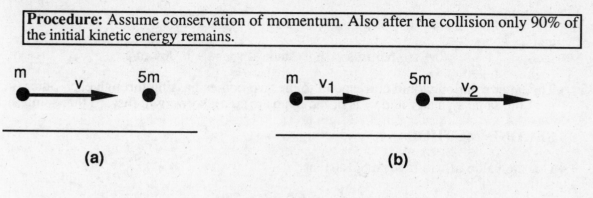

Figure SG 8.2: a). Two gliders before collision.
b). Two gliders after collision.

Since we are told that these bodies are air track gliders, we will assume that the collision conserves momentum. The arrangement before and after collision is shown in Figure SG 8.2 which also defines the symbols used. We choose the initial direction to be positive and let:

Symbol	Quantity Represented
m	mass of first glider $= 100$ g
v	initial velocity of first glider $= 6.0$ m/s
v_1	velocity of first glider after collision
v_2	velocity of second glider after collision

Momentum conservation gives:

$$m\,v \; = \; m\,v_1 + 5\,m\,v_2, \qquad\qquad \text{(SG 8.1)}$$

while energy conservation gives:

$$\tfrac{1}{2} m\,v^2 \; = \; 0.9 \times [\tfrac{1}{2} m\,v_1{}^2 + \tfrac{1}{2}(5\,m)\,v_2{}^2]. \qquad \text{(SG 8.2)}$$

Since m is a common factor in all terms, we can rewrite the equations as

$$v \; = \; v_1 + 5\,v_2 \qquad\qquad \text{(SG 8.1')}$$

and

$$v^2 \; = \; 0.9\,v_1{}^2 + 4.5\,v_2{}^2. \qquad\qquad \text{(SG 8.2')}$$

FIRST METHOD

Using SG 8.1' we can eliminate v from SG 8.2', to obtain

$$20.5\,v_2{}^2 + 10\,v_1 v_2 + 0.1\,v_1{}^2 \; = \; 0.$$

This quadratic has two solutions

$$v_2 \; = \; -0.478\,v_1 \qquad \text{and} \qquad v_2 \; = \; -0.0102\,v_1.$$

Since we know v, we can use SG 8.1' to find the two possible sets of v_1 and v_2

$$v_1 = -4.3 \text{ m/s} \qquad \text{and} \qquad v_2 = 2.1 \text{ m/s}$$

and

$$v_1 = 6.3 \text{ m/s} \qquad \text{and} \qquad v_2 = -0.064 \text{ m/s}.$$

The latter solution would correspond to the two bodies passing through each other (v_1 is of the same sign as v and v_2 is of the opposite sign), so we will discard that solution.

SECOND METHOD

Use the value of v to rewrite SG 8.1' as

$$v_1 = 6.0 \text{ m/s} - 5\, v_2.$$

substitute this into SG 8.2' giving

$$36 \text{ m}^2/\text{s}^2 = 0.9(6.0 \text{ m/s} - 5v_2)^2 + 4.5\, v_2^2$$

$$= 32.4 \text{ m}^2/\text{s}^2 - (27 \text{ m/s})v_2 + 18\, v_2^2 ,$$

or

$$18v_2^2 - (27 \text{ m/s})v_2 - 3.6 \text{ m}^2/\text{s}^2 = 0.$$

The two roots are 2.1 m/s and –0.064 m/s as before. Then SG 8.1' can be used to find the corresponding values for v_1 as before.

A corresponding argument could be made by solving SG 8.1' for v_2 and using that in SG 8.2' to find the pairs of v_1.

From any of these approaches we conclude that the solutions are that the first glider reverses its direction and has a speed of 4.3 m/s while the second glider moves at a speed of 2.1 m/s in the direction that the first glider was originally traveling.

PROBLEM SG 8.3: After suffering an ellastic collision with a stationary puck of the same mass, an air hockey puck is observed to have a momentum whose magnitude is only 1/3 the magnitude of the initial momentum. What angle does it make with its initial direction.

SOLUTION:

Procedure: Since these are air pucks, we will consider this is a two dimensional scattering problem without friction. Write the equations showing conservation of each component of momentum and an equation showing conservation of energy. Solve for the unknown angle.

We sketch the problem in Figure SG 8.3. The symbols used are:

SYMBOL	MEANING
m	mass of either puck
v	initial velocity of puck A
v_1	final velocity of puck A
v_2	final velocity of puck B

Figure SG 8.3:   a). Before the collision. b). After the collision.

We orient our coordinates so the collision occurs in the x-y plane and choose x and y as shown in Figure SG 8.3a. In terms of the symbols adopted this means that

$$|m\, v_1| = |\tfrac{1}{3} m\, v|$$

or that

$$v_1 = \frac{1}{3} v.$$

Using the angles shown in Figure SG 8.3b, momentum conservation gives:

and

$$(x\text{-component}) \qquad m\, v = m\, v_1 \cos \theta + m\, v_2 \cos \phi \qquad \text{(SG 8.4)}$$

$$(y\text{-component}) \qquad m\, v_1 \sin \theta = m\, v_2 \sin \phi , \qquad \text{(SG 8.5)}$$

where θ and ϕ are the angles that pucks A and B make with the x-axis. Energy conservation gives:

$$\frac{1}{2} m\, v^2 = \frac{1}{2} m\, v_1^2 + \frac{1}{2} m\, v_2^2. \qquad \text{(SG 8.6)}$$

Divide out the common factor of m and use $v_1 = \frac{1}{3} v$ to get

$$v = \frac{1}{3} v \cos \theta + v_2 \cos \phi , \qquad \text{(SG 8.4')}$$

$$\frac{1}{3} v \sin \theta = v_2 \sin \phi , \qquad \text{(SG 8.5')}$$

and

$$\frac{8}{9} v^2 = v_2^2. \qquad \text{(SG 8.6')}$$

Rewrite Equation SG 8.4' as

$$v_2 \cos \phi = v \left(1 - \frac{1}{3} \cos \theta \right)$$

Square this and add the result to the square of Equation SG 8.5' and eliminate ϕ by using the trig identity

$$\sin^2 \phi + \cos^2 \phi = 1$$

to get

$$v_2{}^2 = \left(\frac{10}{9} - \frac{2}{3} \cos \theta \right) v^2.$$

Substituting this into the equation SG 8.6' gives

$$\frac{8}{9} v^2 = \left(\frac{10}{9} - \frac{2}{3} \cos\theta \right) v^2,$$

so

$$\cos \theta = \frac{1}{3}$$

or

$$\theta = 71°.$$

The puck is deflected by 71° from its original direction.

PROBLEM SG 8.4: A physics teacher stands on a freely rotating platform. She holds a dumbbell in each hand of her outstretched arms while a student gives her a push until her angular velocity reaches 2.0 rad/s. When the freely spinning professor pulls her hands in close to her body, her angular velocity increases to 7.0 rad/s. What is the ratio of her final kinetic energy to her initial kinetic energy? Where does this energy originate?

SOLUTION:

> **Procedure:** The term freely spinning suggests that friction can be neglected. Thus the total energy and the total angular momentum should be conserved. There is no translation so we need not consider linear momentum. Write the expression for angular momentum and conserve the angular momentum when the arms are moved towards her body.

The situation is shown in Figure SG 8.4 with a) showing the initial arrangement of arms out and angular speed $\omega_0 = 2.0$ rad/s and b) showing the final arrangement with arms in and angular speed $\omega_1 = 7.0$ rad/s.

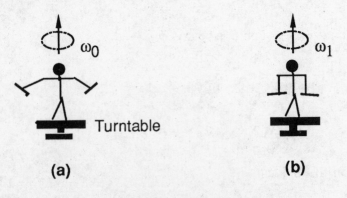

Figure SG 8.4:   a). Starting with arms out.and speed ω_0.
b). with arms in and speed ω_1.

Since the platform is freely rotating, we assume momentum conservation. Let I_i be the initial moment of inertia (arms out) and I_f be the final moment of inertia (arms in), then:

$$I_i \, \omega_0 = I_f \, \omega_1$$

and so

$$I_f = I_i \frac{\omega_0}{\omega_1}.$$

The initial kinetic energy is

$$KE_i = \frac{1}{2} I_i \, \omega_0^2,$$

while the final kinetic energy is

$$KE_f = \frac{1}{2} I_f \, \omega_1^2.$$

This gives:

$$\frac{KE_f}{KE_i} = \frac{I_f}{I_i} \frac{\omega_1^2}{\omega_0^2} = \frac{\omega_1}{\omega_0} = 3.5.$$

So momentum is conserved but kinetic energy increases. The kinetic energy must come from the work done by the teacher in moving her arms in to her sides. That is, food energy has been converted to work which was converted to kinetic energy.

PROBLEM SG 8.5: A driver in an automobile traveling at 35 miles per hour sees a child in the road. Assuming the frictional coefficient is 0.72 for this road and that the car is 50 ft away from the child when the brakes are applied, will the child be hit? How much time elapses before the car stops?

SOLUTION:

Procedure: Use stopping distance formula or kinematical laws to calculate stopping distance and time.

We can use the results of Problem 8.23 of the text to quickly get the first answer. If s is the stopping distance in feet, then

$$s(\text{ft}) = \frac{[v(\text{mph})]^2}{30\mu},$$

where v is the speed in mph and μ is the coefficient of friction. Using $\mu = 0.72$ and $v = 35$ mph, we get $s = 57$ ft, so the child will be hit.

This formula for stopping distance follows from the basic kinematical laws. The forces acting are shown in Figure SG 8.5 which is a free body diagram for the car.

$$\text{The frictional force} = -\mu N = -\mu m g = ma$$

where a is the acceleration and N is the normal force. Then

$$a = -\mu g.$$

The law of motion relating velocity and acceleration is

$$v^2 = v_0^2 + 2\,a\,x,$$

Setting $v = 0$ gives x as the stopping distance, so

$$\text{stopping distance} = \frac{v_0^2}{\mu g}.$$

Choosing English units gives the stopping distance formula used.

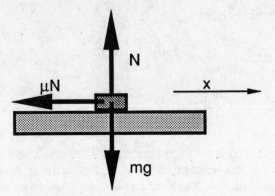

Figure SG 8.5: Free body force diagram for automobile sliding with locked wheels.

To answer the second part of the question, use another kinematical law:

$$v = v_0 + a\,t.$$

Setting $v = 0$ with the a above gives us the stopping time,

$$t = \frac{-v_0}{a} = \frac{v_0}{\mu\,g},$$

where

$$v_0 = 35 \text{ mph} = 35 \times \frac{1 \text{ hr}}{3600 \text{ s}} \times \frac{5280 \text{ ft}}{\text{mile}} = 51 \text{ ft/s}.$$

and

$$g = 32 \text{ ft/s}^2.$$

The stopping time is

$$t = \frac{51 \text{ ft/s}}{0.72 \times 32 \text{ ft/s}^2} = 2.2 \text{ s}.$$

PROBLEM SG 8.6: A meteor is detected heading towards the earth on a path which is nearly a straight head on collision. The meteor was detected at a distance of seven times the radius of the earth ($R_0 = 7 \times R_e$) and a relative speed of 1.0×10^4 m/s. What will be the velocity when the meteor reaches the top of the atmosphere ($R \approx R_e$)? Why can't we easily answer the question, how much time will elapse between detection and collision?

SOLUTION:

> **Procedure:** Use conservation of energy.

$$\text{total energy} = \text{Kinetic} + \text{Potential} = \frac{1}{2}m\,v^2 - \frac{G\,m}{R}.$$

GIVEN:

Quantity	Symbol	Value
initial distance	R_0	$7.0 \times R_e$
initial velocity	v_0	1.0×10^4 m/s
final distance	R_e	
final velocity	v_f	?

$7\,R_e$

R_e

Figure SG 8.6:   Motion diagram for meteor falling on the earth.

This gives

$$\frac{1}{2}m\,v_0^2 - \frac{G\,m}{7\,R_e} = \frac{1}{2}m\,v_f^2 - \frac{G\,m}{R_e},$$

so

$$v_f^2 = v_i^2 + \frac{6\,G}{7\,R_e} = v_0^2 + \frac{6}{7}g\,R_e$$

$$= (10^4 \text{ m/s})^2 + \frac{6}{7}9.8 \text{ m/s}^2 \times 6.38 \times 10^6 \text{ m}$$

$$= 1.54 \times 10^8 \text{ m}^2/\text{s}^2,$$

$$v_f = 1.2 \times 10^4 \text{ m/s}.$$

We can not easily answer the question about the elapsed time since the acceleration is not constant, also frictional forces will be important as the meteor passes through the atmosphere. Either advanced mathematics or graphical analysis will be needed.

PROBLEM SG 8.7: What is the period of a geostationary earth satellite. By what factor would this period need to be increased in order for the satellite to escape?

SOLUTION:

> **Procedure:** A satellite can escape only if its total energy is non-negative; thus the critical escape condition is total energy (kinetic + potential) equal to zero.

A geostationary earth satellite has a 24 hr period in order to keep over the same location. Hence its velocity must satisfy:

$$\text{period} \quad = P = \frac{2\pi R}{v}$$

$$= 24 \, \text{hr} \frac{3600 \, \text{s}}{\text{hr}} = 8.784 \times 10^5 \, \text{s}$$

where R is the orbital radius and v is the velocity in the orbit.

For a circular orbit the centripetal force equals the gravitational force or

$$\frac{m \, v^2}{R} = \frac{G \, M \, m}{R^2}$$

where M is the mass of the earth and m is the mass of the satellite. We can use this last equation to find v, giving

$$v = \sqrt{\frac{GM}{R}}$$

If the satellite had escape velocity, the total energy (kinetic plus potential) must be zero (definition of escape velocity), so the minimum velocity for escape at the same R would satisfy

$$\frac{1}{2} m \, v_{esc}^2 = \frac{GmM}{R}, \quad \text{(kinetic = -- potential)}$$

which is

$$v_{esc} = \sqrt{\frac{2GM}{R}} \, .$$

So the ratio is

$$\frac{v_{esc}}{v} = \sqrt{2}$$

independent of R and v.

PROBLEM SG 8.8: An air puck of mass m, which was moving at 50 cm/s, strikes a stationary puck of mass 2m. The collision is perfectly elastic. The first puck moves off at an angle of 60° with respect to its initial direction. What is the velocity of the initial puck after the collision?

SOLUTION:

> **Procedure:** Write the momentum conservation laws component by component and the conservation law of energy and solve for the final velocity.

We can use Figure SG 8.2 provided we choose

$$
\begin{aligned}
m_B &= 2m_A = 2m \\
v &= 50 \, \text{cm/s} \\
\theta &= \text{final angle of puck } A = 60° \\
\phi &= \text{final angle of puck } B.
\end{aligned}
$$

The x-component of momentum conservation is

$$mv = mv_1 \cos \theta + 2mv_2 \cos \phi,$$

for the y-component $(p_{Ay} + p_{By}) = 0$

$$0 = mv_1 \sin \theta - 2mv_2 \sin \phi,$$

and energy conservation gives

$$\tfrac{1}{2}mv^2 = \tfrac{1}{2}mv_1{}^2 + \tfrac{1}{2}2mv_2{}^2.$$

The most direct way to find v_1, is to solve the first two equations for v_2 and use the result to find v_1 with the third equation; that is, solve the x-momentum and the y-momentum equations as

$$v_2 \cos \phi = \frac{1}{2}(v - v_1 \cos \theta)$$

and

$$v_2 \sin \phi = \frac{1}{2}v_1 \sin \theta.$$

Squaring and adding these gives us

$$v_2{}^2 = \tfrac{1}{4}\left(v^2 - 2vv_1 \cos \theta + v_1{}^2\right).$$

Then energy conservation gives

$$v^2 = v_1{}^2 + 2v_2{}^2 = v_1{}^2 + \tfrac{1}{2}\left\{ v^2 - 2vv_1 \cos \theta + v_1{}^2 \right\}$$

$$= \tfrac{3}{2}v_1{}^2 + \tfrac{1}{2}v^2 - vv_1 \cos \theta,$$

which, rewritten, is

$$v_1{}^2 - \tfrac{2}{3} vv_1 \cos \theta - \tfrac{1}{3}v^2 = 0.$$

The solution of this is

$$v_1 = \frac{v}{3}\left(\cos \theta \pm \sqrt{3 + \cos^2 \theta}\right)$$

and, since $\cos 60° = \frac{1}{2}$,

$$1 = \frac{v}{3}\left(\frac{1}{2} \pm \sqrt{\tfrac{13}{4}}\right) = \frac{v}{6}(1 \pm 3.6),$$

so

$$v_1 = 0.77\, v \qquad \text{or} \qquad v_1 = -0.43\, v.$$

If we assume that the scattering was forward as shown in Figure SG 8.3, we conclude that

$$v_1 = 0.77 \, v = 39 \text{ cm/s}.$$

The other root corresponds to $\theta = 120°$ and comes from

$$\cos 120° = -\cos 60°.$$

PROBLEM SG 8.9: A cylinder of mass 18 kg and radius 75 cm rolls down an inclined plane whose slope is 30° from a height of 1.5 m. Assume the cylinder has an initial horizontal velocity of 2.0 m/s and that it rolls without slipping. What is the translational speed at the bottom of the ramp?

SOLUTION:

Procedure: Use conservation of energy. Since there is no slipping, the angular velocity of the cylinder is related to its translational velocity.

A sketch of the problem is shown in Figure SG 8.7.

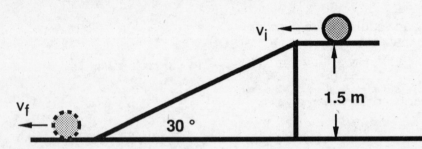

Figure SG 8.7:   Cylinder rolling, without slipping, down an inclined plane.

We will use the following symbols

Quantity	Symbol	Value
initial translational velocity	v_i	2.0 m/s
initial rotational velocity	ω_i	
final translational velocity	v_f	
final rotational velocity	ω_f	
height of inclined plane	h	1.5 m
angle of inclination	θ	30°
radius of cylinder	r	0.75 m
mass of cylinder	m	1.8 kg

The relation between the rotational speed and the translational speed is given by the requirement that there is no slipping. This gives

$$\omega r = v \qquad \text{or} \qquad \omega = \frac{v}{r}.$$

Choose the zero of potential energy at the bottom of the plane and let I be the moment of inertia of the cylinder about the contact point, let the subscript $_f$ denote the final values, and the subscript $_i$ denote the initial values. Note that the rotational frequency of the cylinder about the contact point is the same as its rotational frequency about its center. Energy conservation requires

(Kinetic energy + Potential energy)$_{initial}$ =

(Kinetic energy + Potential energy)$_{final}$,

so

$$\tfrac{1}{2} m \, v_i{}^2 + \tfrac{1}{2} I \omega_i{}^2 + mgh = \tfrac{1}{2} m \, v_f{}^2 + \tfrac{1}{2} I \omega_f{}^2 ,$$

where h is the initial height.

Using the relation between ω and r and the value of I from text Table 6.3 gives

$$\tfrac{1}{2} I \omega^2 = \tfrac{1}{2}\left(\tfrac{3}{2} m r^2 \omega^2\right) = \tfrac{3}{4} m r^2 \frac{v^2}{r^2} = \tfrac{3}{4} m v^2 .$$

The energy equation is then

$$\tfrac{1}{2} m \, v_i{}^2 + \tfrac{3}{4} m v_i{}^2 + mgh = \tfrac{1}{2} m \, v_f{}^2 + \tfrac{3}{4} m v_f{}^2 .$$

The result is independent of m and θ,

$$v_f{}^2 = v_i{}^2 + \frac{4}{5} gh .$$

So the translational speed of the cylinder at the bottom of the ramp is 4.0 m/s.

PROBLEM SG 8.10: How much energy is dissipated when a 3.0 kg grinding wheel of radius 0.10 m is brought to rest from an initial velocity of 3000 rpm? What force must be applied tangentially to the rim if this is to be done in 3.0 seconds?

SOLUTION:

Procedure: Since the final state is at rest, the initial kinetic energy is the total energy and the work done must be equal to this energy.

If I is the moment of inertia of a solid wheel and ω_0 is its angular velocity then

$$\text{Kinetic energy} = \tfrac{1}{2} I \omega_0{}^2$$

where

$$I = \tfrac{1}{2} M R^2$$

with M the mass of the disk and R the radius of the disk. For this problem

This gives a kinetic energy of 7.4×10^2 J.

If the applied force is represented by F and the angular acceleration by α, then the torque, τ, is given by

$$\tau = I\alpha \, ,$$

and the angular velocity, ω, by

$$\omega = \omega_0 - \alpha t \, ,$$

where t is the elapsed time and the minus sign shows that the force decelerates the wheel.

This gives for $t_f = 3$ s, since $w(3 \text{ s}) = 0$,

$$F = \frac{I\omega_0}{Rt_f} \, .$$

Entering numbers gives the resulting force.

Chapter 9
Fluids

Important Terms

Fill in the blank with the appropriate word or words. Three important equations or Laws are included in addition to the Important Terms from the list.

1. A substance which cannot maintain its own shape is called a(n) _____.

2. The SI unit of pressure is called the _____. It is equivalent to $1 \frac{N}{m^2}$.

3. Pressure measured relative to the surrounding atmosphere is called _____ pressure.

4. "A body, either submerged or floating, is buoyed up by a force that is equal to the weight of the displaced fluid" is a statement of _____.

5. The _____ is a parameter which tells us if fluid flow is laminar or turbulent.

6. _____ flow is a smooth streamline flow.

7. _____ action, which results from surface tension and adhesive forces between a fluid and the walls of a small tube, shows when the fluid rises in the small tube above the level of the surrounding fluid.

8. The pressure difference between two points in a pipe containing a flowing viscous fluid is given by _____ Law.

9. A(n) _____ is a device for measuring the velocity of a fluid flowing through a pipe.

10. The upward force exerted on a body in a fluid is called the _____ force.

11. A falling body entering a fluid slows its velocity until it reaches _____ velocity when the upward fluid force equals the gravitational force.

12. _____ is the force per unit area. Its SI unit is the pascal (Pa).

13. _____ Law gives us the relation between the drag force exerted by a fluid on a moving body and the velocity of the body.

14. _____ principle tells us that the pressure exerted at any point of an enclosed fluid is transmitted to every part of the fluid.

15. The _____ of a fluid is a measure of the fluid's internal friction.

16. The equation of _____ must hold if the total amount of fluid crossing any surface is constant.

17. The _____ of a fluid is a measure of the force per unit length needed to remove a rigid body from the fluid.

18. _____ equation is the form that the conservation of energy takes for fluids.

19. If neighboring layers of a fluid flow smoothly past each other, the resulting paths are called _____.

20. _____ flow is characterized by irregular, complex motion of the fluid.

Write the definitions of the following:

21. streamline (9.5):

22. Reynolds number (9.9):

23. fluid (9.0):

24. laminar flow (9.5):

25. equation of continuity (9.5):

26. pressure (9.1):

27. terminal velocity (9.7):

28. surface tension (9.4):

29. turbulent flow (9.5):

30. pascal (9.1):

31. Stokes' Law (9.7):

32. Bernoulli's equation (9.5):

33. Poiseuille's law (9.6):

34. Pascal's principle (9.2):

35. viscosity (9.6):

36. Archimedes' principle (9.3):

37. Venturi meter (9.5):

38. buoyant force (9.3):

Answers to 1-20, Important Terms

1. fluid; 2. pascal; 3. gauge; 4. Archimedes' principle; 5. Reynolds number; 6. Laminar; 7. capillary; 8. Poiseuille's; 9. Venturi meter; 10. buoyant; 11. terminal; 12. Pressure; 13. Stokes'; 14. Pascal's; 15. viscosity; 16. continuity; 17. surface tension; 18. Bernoulli's; 19. streamlines; 20 Turbulent

Sample Solutions

PROBLEM SG 9.1: A 200 kg motorized tricycle is supported by three identical tires inflated to a gauge pressure of 200 kPa. Ignoring any possible effects of tread thickness, calculate the area of contact between each tire and the road. Hint: Neglect any buoyant force due to the earth's atmosphere.

SOLUTION:

Procedure: The tricycle is in equilibrium so the upward and downward forces balance. Through a series of action-reaction pairs, the gravitational force is balanced by the force provided by the tires. This latter force is pressure times contact area.

Since the tricycle is in static equilibrium the normal force provided by the surface upon which the tricycle rests is equal to the mass times the acceleration of gravity as shown symbolically in Figure SG 9.1. That is if the mass is m and g is the acceleration of gravity, then

$$N = mg,$$

and the reaction pair is the force provided by the tires. Hence

$$F = mg.$$

Figure SG 9.1: Forces between tricycle and road.

The force, F, provided by the tires will be the pressure in the tires times the surface area of contact. This force must also be the weight of the tricycle, mg.

If we let A be the area of one tire and assume all tires are equal, then

$$F = 3 \times p \times A,$$

where the 3 comes from the three tires and p is the true pressure inside the tires. Equating these gives

$$A = \frac{mg}{3p}.$$

The true pressure is found from the gauge pressure by adding the pressure of one atmosphere

$$p = p_{gauge} + p_{atmosphere} = 200 \text{ kPa} + 101.3 \text{ kPa}$$

$$= 301 \text{ kPa}.$$

Hence

$$A = \frac{200 \text{ kg} \times 9.80 \text{ m/s}^2}{3 \times (3.01 \times 10^5 \text{ Pa})}$$

$$= \frac{200 \times 9.8}{9.03 \times 10^5} \times \frac{\text{kg m/s}^2}{\text{Pa}}$$

$$= 2.17 \times 10^{-3} \times \frac{\text{kg m/s}^2}{\text{Pa}}$$

$$= 2.17 \times 10^{-3} \times \frac{\text{N}}{\text{N/m}^2} = 2.17 \times 10^{-3} \text{ m}^2.$$

Each tire has a contact area of about 22 cm^2.

PROBLEM SG 9.2: A wooden board 1.0 m × 15 cm × 2.0 cm has a density of 0.50 g/cm^3. The board is floated in water and lead is placed on top of the board. How much lead can be placed on top of the board and have the top of the board level with the water? Assume the experiment is done at 20°C.

SOLUTION:

Procedure: Use Archimedes' Principle. Calculate the volume (and hence the mass) of water displaced by the board and set this equal to the mass of the board plus the lead.

This problem can be solved by a straightforward application of Archimedes' principle. The board plus lead will displace a mass of water equal to the combined mass of the board plus lead. Since the board floats level with the water, as seen in Figure SG 9.2, the volume of water displaced is the same as the volume of the board.

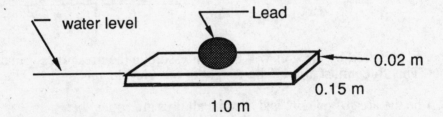

Figure SG 9.2:   Lump of lead on a wooden board which is floating on water.

LET:

Symbol	Quantity	Value
m_{Pb}	mass of lead	
m_b	mass of board	
m_w	mass of displaced water	
V_b	volume of the board	
V_w	volume of displaced water	
ρ_b	density of wood	0.50 g/cm^3
ρ_w	density of water	998 kg/m^3

then

$$m_w = m_b + m_{Pb}.$$

and

$$V_w = V_b,$$

with

$$m_w = V_w \times \rho_w$$

and

$$m_b = V_b \times \rho_b.$$

Thus

$$m_{Pb} = m_w - m_b = V_w \times \rho_w - V_b \times \rho_b.$$

$$= V_b (\rho_w - \rho_b).$$

We can now enter the numbers

$$V_b = 1.0 \text{ m} \times 0.15 \text{ m} \times 0.02 \text{ m} = 3.0 \times 10^{-3} \text{ m}^3$$

and using the numbers in Table SG 9.2 (convert the density of wood to kg/m^3) we find that the lead has a mass of

$$1.5 \text{ kg.}$$

PROBLEM SG 9.3: A container filled partway with water has a total mass of 2.50 kg (container plus water). A piece of lead with a mass of 1.13 kg is suspended from a spring scale. The scale is calibrated to read weight. If the scale is arranged such that the lead is completely submerged in the water. What is the reading on the spring scale? Assume the experiment is carried out at 20°C.

SOLUTION:

Procedure: Use Archimedes' principle. The water will exert an upwards force equal to the weight of the displaced water.

Figure SG 9.3: Free body diagram for the lead held in water. The buoyant force is denoted by F_w while the tension acting on the spring is T.

Archimedes' principle tells us that the water exerts an upwards force equal to the weight of the displaced water. The spring scale should read the weight of the lead less this upward force. This is expected since the tension of the lower end of the spring scale is the sum of the gravitational force on the mass and the buoyant force of the water. That is, in terms of the free body diagram for the lead as shown in Figure SG 9.3,

$$T = mg - F_w$$

where m is the mass of the lead, g the acceleration of gravity, F_w is the buoyant force, and, T is the tension applied to the scale and hence the weight read by the spring scale.

To calculate the weight of the displaced water we need to find the volume of lead.

$$\text{Volume of lead} = \frac{\text{mass of lead}}{\text{density of lead}}$$

$$= \frac{1.13 \text{ kg}}{1.13 \times 10^4 \text{ kg/m}^3}$$

$$= 1.00 \times 10^{-4} \text{ m}^3.$$

This corresponds to a weight of water of

$$F_w = \text{density} \times \text{volume} \times \text{acceleration of gravity}$$

$$= 998 \text{ kg/m}^3 \times 1.00 \times 10^{-4} \text{ m}^3 \times 9.80 \text{ m/s}^2$$

$$= 0.978 \text{ N}.$$

The spring scale will read

$$T = 1.13 \text{ kg} \times 9.8 \text{ m/s}^2 - 0.978 \text{ N}$$

$$= 10.1 \text{ N}.$$

PROBLEM SG 9.4: A 25.0 cm high by 20 cm diameter can, resting on a table, is filled to the brim with SAE 10 motor oil (density = 0.95 g/cm^3). 20.0 cm from the top of the can is a 10.0 cm long horizontal tube with a cross-section of 0.20 cm^2. How far from the base of the can will the oil hit the table?

SOLUTION:

> **Procedure:** Use Bernoulli's equation to calculate the pressure at the depth of the fluid. Use Poiseuille's law to calculate the flow rate and hence the exit velocity. Newton's laws will then determine the distance of travel while falling the extra 5.0 cm.

Since the diameter of the can is large when compared to the diameter of the tube, it is reasonable to neglect the motion of the oil except when it flows through the tube and leaves the can. We can then use Bernoulli's equation to calculate the pressure in the can at the level of the tube. We take the pressure at the surface of the oil to be zero — i.e.. neglect air pressure. Then the oil pressure at the level of the tube is

$$\text{pressure} = p_i = \rho gh,$$

where

p_i is the interior pressure at the level of the tube, ρ is the oil density, g is the acceleration of gravity and h is the 20 cm depth of the oil at the tube.

Since the pressure of the exit of the tube is nearly zero, we can use Poiseuille's law to find the flow rate through the tube, that is

$$p_i - p_e = p_i = 8\frac{Q\eta L}{\pi R^4},$$

where

p_e is the exit pressure, Q is the flow rate, η is the viscosity, L is the tube length, R is the tube radius and SI units must be used. Solving this for the flow rate gives

$$Q = \frac{\pi R^4 p_i}{8\eta L}.$$

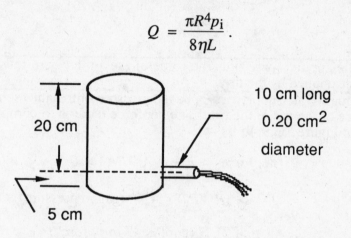

20 cm

5 cm

10 cm long

0.20 cm^2

diameter

Figure SG 9.4: Oil flowing out of a tube in the side of a can.

We can convert this flow rate to an exit velocity, v_e by

$$v_e = \frac{Q}{\text{tube area}} = \frac{Q}{\pi R^2}.$$

Thus

$$v_e = \frac{p_i R^2}{8\eta L} = \frac{\rho g h R^2}{8\eta L}.$$

The vertical and horizontal motions of the fluid will be assumed independent so a fluid element should hit the table in a time given by

$$h_o = \frac{1}{2} g \, t_f^2,$$

where h_o is the height above the table and t_f is the time it takes to hit the table. Then the horizontal distance traveled is

$$\text{distance} = v_e \, t_f = \frac{\rho g h R^2}{8\eta L}\sqrt{\frac{2h_o}{g}}$$

$$= \frac{\rho h R^2 \sqrt{2g h_o}}{8\eta L}.$$

Using SI units, R^2 = area/π, and text Table 9.4 for the viscosity, we find

$$\text{distance} = 0.0075 \text{ m} = 0.75 \text{ cm}.$$

This is the horizontal distance from the end of the tube, so the total distance is 10.75 cm.

PROBLEM SG 9.5: A bicycle pump has a piston of diameter 2.54 cm (1 inch). What force must be exerted on the piston to add air to a tire at a gauge pressure of 413 kPa (60 lb/in^2)?

SOLUTION:

> **Procedure:** The applied force plus the force of the atmosphere must be greater than the internal force applied by the tire.

Let F be the force we are supplying, P_a be the pressure of the atmosphere, and P be the internal pressure of the gas in the tire. The net force tending to compress the air in the tire as shown in Figure SG 9.4b is

$$\text{net force} = F + P_aA - PA > 0.$$

Hence

$$F > (P - P_a) \times A = (\text{gauge pressure}) \times A$$

$$> 21 \text{ N (approx 47 lb of force)}$$

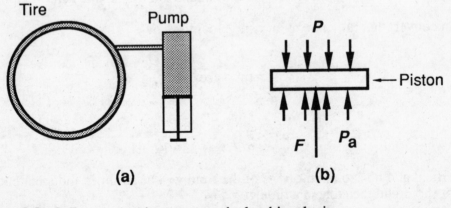

(a) (b)

Figure SG 9.5: a). Air pump attached to bicycle tire.
b). Forces on air pump piston

Chapter 10
Thermal Physics

Important Terms

Fill in the blanks with the appropriate word or words.

1. The _____ temperature scale has the steam point at 212 degrees.

2. Two objects at the same temperature are said to be in thermal _____.

3. The _____ temperature scale divides the interval between the ice point and the steam point into 100 intervals and takes the ice point as its zero.

4. If a material undergoes a change of phase, the amount of heat absorbed or emitted per unit mass is called the _____ .

5. The ratio of the thickness of a material to its thermal conductivity is called the material's _____ .

6. _____ is the minimum possible temperature.

7. The quantity which relates the size of an object at a given temperature to its size at another temperature is called the thermal _____ .

8. An iron poker has one end in a fire. After a period of time, the other end also gets hot. The heat has been transferred by the process of _____ .

9. The heat from the sun reaches the earth by _____ .

10. The _____ of an object is a number assigned to indicate the heat contained in the object.

11. The basic divisions of all currently used temperature scales is called the _____ .

12. _____ are devices used to measure temperatures.

13. The _____ temperature scale uses absolute zero as its reference zero.

14. The heat required to raise the temperature of one gram of water one degree Celsius is called the _____ .

15. Hot air rises and carries off heat from a hot body. This process is called _____ .

16. The ratio of the amount of heat required to raise the temperature of a body 1 C° to the mass of the body is called the _____ of the body.

17. For small temperature changes, we assume that one dimensional thermal behavior is _____ thermal expansion.

18. The _____ is the energy absorbed or emitted as a substance changes from a solid to a liquid or conversely.

19. The _____ is the energy absorbed or emitted as a substance changes from a liquid to a vapor or conversely.

20. The _____ is a measure of the ability of a substance to transfer heat by conduction from a boundary at one temperature to a boundary at another.

Write the definitions to the following:

21. Kelvin temperature scale (10.2):

22. latent heat of transformation (latent heat) (10.6):

23. temperature (10.1):

24. thermometer (10.1):

25. heat capacity (10.5):

26. specific heat (10.5):

27. convection (10.7):

28. radiation (10.7):

29. conduction (10.7):

30. R value (10.7):

31. Fahrenheit temperature scale (10.2):

32. absolute zero (10.2):

33. linear thermal expansion (10.3):

34. thermal conductivity (10.7):

35. calorie (10.4):

36. thermal equilibrium (10.1):

37. thermal expansion coefficient (10.3):

38. Celsius temperature scale (10.2):

39. heat of fusion (10.6):

40. heat of vaporization (10.6):

Answers to 1-20, Important Terms

1. Fahrenheit; 2. equilibrium; 3. Celsius; 4. latent heat of transformation (latent heat); 5. R value; 6. absolute zero; 7. expansion coefficient; 8. conduction; 9. radiation; 10. temperature; 11. degree; 12. thermometers; 13. Kelvin; 14. calorie; 15. convection; 16. specific heat capacity (specific heat); 17. linear; 18. heat of fusion; 19. heat of vaporization; 20. thermal conductivity

General Comments

In mechanics you were urged to draw motion plots or force diagrams. The analogous plots for thermal physics is to draw heat flow diagrams. Large bodies or reservoirs of heat are indicated by their temperature. Heat flow and work can be indicated by arrows. An example is shown in Figure SG 10.1 where heat Q_1 is extracted at temperature T_1, heat Q_2 is exhausted at temperature T_2 while work W is done. Figures such as SG 10.1 not only help your thinking, but they also make the meaning of the variables more obvious.

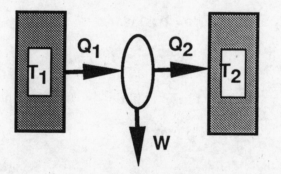

Figure SG 10.1:   An example of a heat flow diagram.

Sample Solutions

PROBLEM SG 10.1: A concrete highway in Alaska consists of slabs 12 m in length. How wide must the expansion joints be to allow for thermal changes over the temperature range from –30 °C to 30 °C? Assume the concrete is of a special type which has a linear expansion coefficient of 9×10^{-6}/C°.

SOLUTION:

> **Procedure:** Calculate the linear expansion of each slab over the temperature range. Investigate the effect of choosing reference length at different temperatures.

The length of the slab should be given by Equation 10.4 of the text:

$$L = L_0 [1 + \alpha(T - T_0)],$$

where L is the length at temperature T, L_0 is the length at temperature T_0, and α is the linear expansion coefficient.

We are given that $\alpha = 9 \times 10^{-6}/C°$ and we know that the length is about 12 m. What may not be clear from the problem statement is if we should take $L_0 = 12$ m or if L_0 is the length at some intermediate temperature. We will show that the choice of the temperature for L_0 has no effect on the answer.

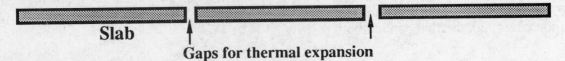

Slab

Gaps for thermal expansion

Figure SG 10.2:   Concrete slabs with gaps for thermal expansion.

FIRST CHOICE OF L_0

Take T_0 to be 30°C and $L_0 = 12$ m. Then for $T = -30°C$, L is given by

$$L = 12 \text{ m } [1 + \frac{9 \times 10^{-6}}{C°}(-30°C - 30°C)]$$

$$= 12 \text{ m}[1 + 0.00054] .$$

So the change in length

$$\Delta L = L - L_0,$$

is given by

$$\Delta L = 12 \text{ m} \times 0.00054 = 0.00648 \text{ m}$$

$$= 6.5 \text{ mm}.$$

SECOND CHOICE OF L_0.

Take

$$T_0 = -30°C$$

and

$$L_0 = 12 \text{ m}.$$

Calculate ΔL for 30°C. This gives the same number.

THIRD CHOICE FOR L_0.

Take

$$T_0 = -30°C$$

and

$$L \text{ at 30°C to be 12 m,}$$

so

$$\Delta L = 12 \text{ m} - L_0.$$

This gives us the following two equations:

$$L(-30°C) = L_0$$

$$L(30°C) = 12 \text{ m}$$

$$= L_0[\ 1 + \alpha(30°C - \{-30°C\})]$$

$$= 1.00054 \times L_0.$$

The latter equation gives us

$$L_0 = \frac{12 \text{ m}}{1.00054},$$

so

$$\Delta L = 12 \text{ m} \frac{1.00054 - 1}{1.00054} = \frac{6.5 \text{ mm}}{1.00054}.$$

Considering the significant figures known, this answer is the same as in the first two approaches. What this result suggests is that if it is not clear which length is L_0, then it probably does not matter which temperature is assigned the value of L_0 as long as you are consistent.

All three approaches give us essentially the same answer.

PROBLEM SG 10.2: Three hundred grams of lead shot is placed in a 2.0 m long wooden tube which is capped on both ends. If the tube is in a vertical position and then quickly inverted, the shot will fall the length of the tube. If this is done 100 times in succession, what will be the increase of the temperature of the shot?

SOLUTION:

Procedure: The work done in inverting the tube is put into potential energy of the shot. The shot falls converting the potential energy to kinetic energy. Upon impact this kinetic energy is converted to heat. Use energy conservation to calculate the kinetic energy which is then completely converted to heat.

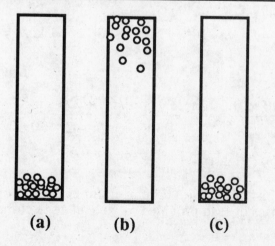

(a)   (b)   (c)

Figure SG 10.3:   a). Tube with shot before inversion.
b). Tube with shot just after inversion and with shot starting to fall.
c). Tube with shot after shot has fallen to bottom of the tube.

The idea behind this problem is illustrated in Figure SG 10.3. In Figure SG 10.3a we have the tube at the beginning of an inversion. The tube is quickly inverted. By "quickly inverted", we mean to make the approximation that none of the shot has a chance to start to fall before the inversion process is completed. Inverting the tube requires work. Part of this work appears as potential energy in the shot. As the shot falls, this potential energy changes to kinetic energy. When the shot hits the bottom of the tube, the kinetic energy is converted into heat. Since the tube is wood, we will assume that none of this heat is lost to the tube.

One additional assumption must be made. We will assume that all of the shot falls through a distance equal to the length of the tube.

GIVEN:

Quantity	Symbol	Value
length of tube	L	2.0 m
mass of shot	m	0.300 kg
number of inversions	n	100
potential energy added per inversion	PE	$m\,g\,L$
heat added to one shot per inversion	ΔQ	PE (in kcal)
specific heat of lead (Table 10.3)	c	0.0305 kcal/kg-C°

If total heat added is $n\,\Delta Q$, then the temperature change ΔT will be given by

$$n\Delta Q = c\,m\Delta T,$$

so

$$\Delta T = \frac{n\,\Delta Q}{mc}.$$

Before we can use the given numbers, we must convert the change in Potential energy to kcal. The expression PE $= mgL$ gives the PE in joules. From table 10.2 of the text, we have

$$1 \text{ joule} = 2.38 \times 10^{-4} \text{ kcal} \qquad \text{or} \qquad 1 \text{ kcal} = 4187 \text{ J}$$

Then

$$\Delta T = \frac{100 \times 0.30 \text{ kg} \times 9.8 \text{ m/s}^2 \times 2.0 \text{ m} \times 2.38 \times 10^{-4} \text{ kcal/J}}{0.0305 \text{ kcal/kg-C°} \times 0.30 \text{ kg}}$$

$$= 15.29 \text{ C°} = 15 \text{ C°}$$

One hundred inversions will raise the temperature of the shot 15 C°. This result is independent of the mass of lead provided as we can neglect the heat absorbed by the wood.

PROBLEM SG 10.3: A cup of negligible heat capacity and heat conductivity holds 210 g of hot water at 90°C. A 200 g block of glass at 27°C is immersed in the water and comes to equilibrium with the water. What is the final temperature of the block of glass?

SOLUTION:

Procedure: Let the heat lost by the block of glass equal the heat gained by the water. Since the water is at a higher temperature and the glass at a lower temperature, we expect the final temperature to be between the two temperatures. Since all temperatures are below or above phase changes, we will not have to worry about heats of transformation. Our basic assumptions are that the heat lost by the water is the heat gained by the glass as shown in Figure SG 10.5c and that, in equilibrium, both are at the same temperature. We can get the needed heat capacities (specific heats) from Table 10.3 of the text.

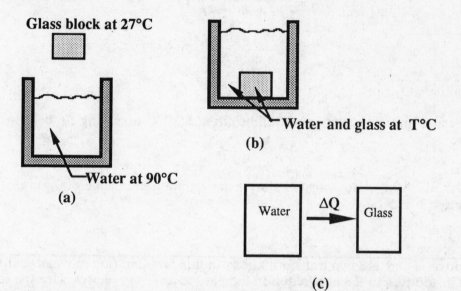

Figure SG 10.4   a). System of water and glass before contact.
b). System of water and glass in contact.
c). Heat flow from water to glass.

GIVEN:

Quantity	Symbol	Value
heat capacity of water	c_W	1.00 cal/g-C°
heat capacity of glass	c_G	0.199 cal/g-C°
mass of water	m	210 g
mass of glass	M	200 g
Initial temperature of water	T_W	90°C
Initial temperature of glass	T_G	27°C
Final temperature of both	T	unknown

LAW TO USE:

Heat loss of water $(-\Delta Q_W)$ = Heat gain of glass (ΔQ_G).

Note that we express the quantities ΔQ as the appropriate quantities for ΔQ positive if there is a heat gain. That is, a negative heat gain is a heat loss. Then,

$$\Delta Q_W = c_W \times m \times (T - T_W) = -c_W \times m \times (T_W - T),$$

and

$$\Delta Q_G = c_G \times M \times (T - T_G).$$

Equating gives

$$c_W \times m \times (T_W - T) = c_G \times M \times (T - T_G)$$

or

$$T = \frac{c_W \; m \; T_W + c_G \; M \; T_G}{c_W \; m + c_G \; M}$$

$$= \frac{19974 \; cal}{249.8 \; cal/C^\circ} = 79.96 \, ^\circ C = 80.0 \, ^\circ C$$

to 3 significant figures.

So the water and glass come to equilibrium at 80.0°C assuming no heat loss to the container.

PROBLEM SG 10.4: How many calories are required to change 800 grams of ice at −20°C to steam at 135°C?

SOLUTION:

> **Procedure:** There are two transformations in this problem (ice to water and water to steam) in addition to the heat required to heat the different phases. Use the tabulated heat capacities to calculate the energy required for each step.

This process can be broken into 5 distinct physical steps:

1. Raise the temperature of the ice from –20°C to 0°C.
2. Melt the ice (temperature remains constant).
3. Raise the temperature of the water from 0°C to 100°C.
4. Change the water to steam (temperature remains constant).
5. Raise the temperature of the steam from 100°C to 135°C.

We will let the symbols ΔQ_1, ΔQ_2, ΔQ_3, ΔQ_4, and ΔQ_5 represent the heat that must be added in each step as shown in Figure SG 10.5.

For steps 1, 3, and 5, $\quad\quad\quad \Delta Q_i = c \times m \times \Delta T \quad\quad$ (Equation 10.6 of text)

where c is the appropriate specific heat, m is the mass, and ΔT is the temperature change.

For steps 2 and 4, $\quad\quad\quad\quad\quad \Delta Q_i = L \times m \quad\quad$ (Equation 10.7 of text)

where L is the appropriate latent heat of transformation and m is the mass.

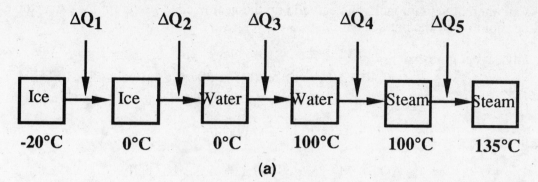

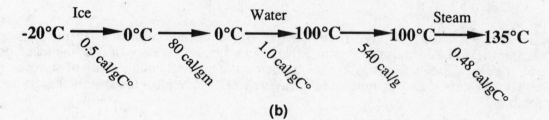

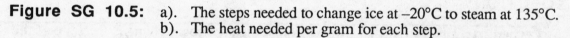

Figure SG 10.5: a). The steps needed to change ice at –20°C to steam at 135°C.
b). The heat needed per gram for each step.

GIVEN:

Quantity	Symbol	Value
specific heat of ice	c_I	0.50 cal/gC°
specific heat of water	c_W	1.00 cal/gC°
specific heat of steam	c_S	0.48 cal/gC°
heat of fusion of water	L_I	80 cal/g
heat of vaporization of water	L_S	540 cal/g
mass (constant for all steps)	m	800 g

Then, for each step, we have:

$$\Delta Q_1 = m \times c_I \times (0°C - [-20°C]) = 8{,}000 \text{ cal}$$

$$\Delta Q_2 = m \times L_I = 64{,}000 \text{ cal}$$

$$\Delta Q_3 = m \times c_W \times (100°C - 0°C) = 80{,}000 \text{ cal}$$

$$\Delta Q_4 = m \times L_S = 432{,}000 \text{ cal}$$

$$\Delta Q_5 = m \times c_S \times (135°C - 100°C) = 13{,}440 \text{ cal}$$

So the total heat needed, ΔQ, is

$$\Delta Q = 597{,}440 \text{ cal} = 6.0 \times 10^5 \text{ cal},$$

since some of the specific heats are given to only two figures.

As an alternative approach, one can add up the heat needed per gram for each step and then multiply by the mass. That is,

total heat per gram =

$$0.5 \text{ cal/gm-C}° \times 20 \text{ C}° + 80 \text{ cal/g} + 1 \text{ cal/gC}° \times 100 \text{ C}°$$
$$+ 540 \text{ cal/g} + 0.48 \text{ cal/gC}° \times 35 \text{ C}°$$

$$= 747 \text{ cal/g}$$

Then

$$\Delta Q = 747 \text{ cal/g} \times 800 \text{ g} = 6.0 \times 10^5 \text{ cal,}$$

as before.

PROBLEM SG 10.5: A styrofoam cooler with a surface area of 0.75 m^2 and an average thickness of 2.5 cm is filled with 2.0 kg of ice at a temperature of 0°C and taken on a fishing trip. How long will it take the ice to melt if the thermal conductivity of styrofoam is 3.2×10^{-4} W/cm-C° and the outside temperature is 35°C? Neglect the specific heat of the styrofoam.

SOLUTION:

> **Procedure:** The energy that flows through the styrofoam changes the state of the ice to water. Assume there is no change in temperature until all the ice is melted.

During the entire time that the ice is in the cooler, heat is being conducted through the outside walls to the inside as shown in Figure SG 10.6a. Since the ice starts at 0°C and we are considering only the time it takes the ice to melt, the interior temperature remains at 0°C. Thus the situation is as illustrated in Figure SG 10.6b.

We are interested in finding how long it takes for enough heat to enter the container to melt 2.0 kg of ice. This amount of heat, ΔQ, is given by

$$\Delta Q = (\text{mass of ice}) \times (\text{heat of fusion of ice})$$

$$= 2000 \text{ g} \times 80 \text{ cal/g} = 160,000 \text{ cal.}$$

The heat flow equation is given by text equation 10.8 as

$$\frac{\Delta Q}{\Delta t} = -K A \frac{\Delta T}{\Delta x} \, ,$$

where ΔQ is the heat that flows in the time interval Δt, K is the thermal conductivity, A is the surface area, and ΔT is the temperature change across thickness Δx. The unknown quantity is Δt, so

$$\Delta t = \frac{\Delta Q}{-K A \dfrac{\Delta T}{\Delta x}} \, .$$

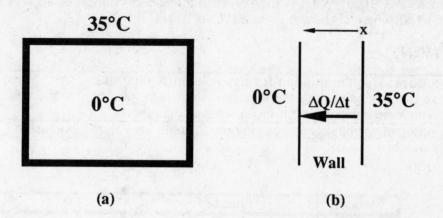

Figure SG 10.6:   a). Cooler with interior at 0°C and exterior at 35°C.
b). Heat flow across wall. Positive direction chosen to make
ΔQ positive.

Before substituting in numbers, we should get all quantities in a consistent set of units. Alternatively, we could work out the units and convert the final answer. Usually the former procedure is simplest. Let us consider each separately:

$\Delta T = 0°C - 35°C = -35C°$ leave as is — notice sign follows from choice of direction in figure SG 10.6b.

$\Delta x = 2.5$ cm   leave as is

$A = 0.75$ m^2 = 7500 cm^2 use cm or convert Δx to meters

$K = 3.2 \times 10^{-4}$ W/cm–C°   consistent with A and Δx in cm

ΔQ = 160,000 cal   convert to W-s or change K

$= 1.6 \times 10^5$ cal $\times$ 4.187 J/cal

$= 6.6992 \times 10^5$ J

So

$$\Delta Q = 6.6992 \times 10^5 \text{ W-s.}$$

Then

$$\Delta t = \frac{6.6992 \times 10^5 \text{ W-s.}}{-3.2 \times 10^{-4} \text{ W/cm-C°} \times 7500 \text{ cm}^2 \frac{-35C°}{2.5 \text{ cm}}}$$

$$= 1.9 \times 10^4 \text{ s} \cong 5.3 \text{ hr.}$$

It will take the ice about $5\frac{1}{3}$ hr to melt

PROBLEM SG 10.6: What is the percent increase in volume of a glass sphere with a radius of 10 cm when it is heated from 20 °C to 100 °C?

SOLUTION:

> **Procedure:** Use the relation for the volume change in the text. Since a percentage change is desired and since the volume change is proportional to the volume, we expect the result to be independent of the volume of the sphere.

This problem can be solved by using the formula for volume expansion (Text Equation 10.5) which is

$$\Delta V = \beta V_0 \Delta T$$

where

Symbol	Quantity
ΔV	change in volume
β	volume coefficient of expansion of glass $= 27 \times 10^{-6}/C°$
V_0	initial volume of glass $(\frac{4}{3} \pi r^3)$
ΔT	change in temperature $= 80$ C°

The percentage change will be

$$\frac{\Delta V}{V_0} \times 100 = 100 \, \beta \, \Delta T = 0.22\%$$

so the volume does not need to be calculated and the percentage change is 0.22%.

Chapter 11
Thermodynamics

Important Terms

Fill in the blanks with the appropriate word or words.

1. An increase of the _____ of a system is a measure of the increase of the disorder of the system.

2. A thermodynamic process in which the volume does not change is called a(n) _____ process.

3. A(n) _____ process is one in which no heat enters or leaves the system.

4. A(n) _____ system interacts with its environment via heat transfer and work.

5. The _____ Law of Thermodynamics states that "Two systems, each in thermal equilibrium with a third, are in thermal equilibrium with each other".

6. A system which requires work to move heat from a lower temperature region to a higher temperature region is called a(n) _____ .

7. Thermodynamic properties of a system, such as volume, temperature and pressure, are called _____ variables.

8. Machines which are supposed to deliver energy from a system and return the system to its initial state without any net heat flow are called perpetual motion machines of the _____ kind.

9. The ratio of the work done to the heat input is called the _____ .

10. _____ is the area of physics concerned with the relationships between heat and work.

11. Conservation of energy is the essential content of the _____ Law of Thermodynamics.

12. The _____ cycle consists of two reversible isothermal and two reversible adiabatic processes.

13. The _____ Law of Thermodynamics tells us that heat cannot flow from a cooler body to a warmer body.

14. A process performed at constant pressure is called a(n) _____ process.

15. If the reversal of controlling factors causes a reversal of the energy transformation, the process is said to be _____ .

16. The ratio of the heat delivered to the work supplied is called the _____ .

17. A(n) _____ process is one in which the temperature of the system does not change.

18. A refrigerator which cools the outside and provides heat to the interior of a building is called a(n) _____.

Write the definitions of the following:

19. isochoric process (11.1):

20. isobaric process (11.1):

21. zeroth law of thermodynamics (11.1):

22. second law of thermodynamics (11.4):

23. adiabatic process (11.1):

24. thermal efficiency (11.2):

25. coefficient of performance (11.3):

26. entropy (11.5):

27. thermodynamic system (11.1):

28. state variable (11.1):

29. reversible process (11.1):

30. perpetual motion of the first kind (11.1):

31. Carnot cycle (11.2):

32. heat pump (11.3):

33. thermodynamics (11.0):

34. first law of thermodynamics (11.1):

35. isothermal process (11.1):

36. refrigerator (11.3):

Answers to 1–18, Important Terms

1. entropy; 2. isochoric; 3. adiabatic; 4. thermodynamic; 5. zeroth; 6. refrigerator; 7. state; 8. first; 9. thermal efficiency; 10. thermodynamics; 11. first; 12. Carnot; 13. second; 14. isobaric; 15. reversible; 16. coefficient of performance; 17. isothermal; 18. heat pump

Modeling

Extensive use of models is made in thermodynamics. We usually do not need to look at the internal structure of the thermodynamic system. The system has a temperature and an internal energy and is able to to work. Heat is transferred into and out of the system by bringing the system into contact with other bodies whose internal structure is often not specified.

Often these external bodies are assumed to be what are called constant temperature reservoirs. Such reservoirs are able to give up heat or absorb heat without changing their temperatures. Such ideal reservoirs do not exist in nature. However, they can be approximated in nature by making the constant temperature body sufficiently large that the internal energy of the reservoir is very much greater than the changes of the internal energy of the system under study. We will make extensive use of this model in this and the next chapter.

Sample Solutions

PROBLEM SG 10.1: Consider a system whose pressure versus volume follows the one shown in Figure SG 11.1a. The system undergoes the transformation $A \Rightarrow B \Rightarrow C$. The total heat taken in is 500 J and the system does 200 J of work. What is the change in the internal energy of the system? What would the change be if the path taken had been $A \Rightarrow D \Rightarrow C$?

SOLUTION:

Procedure: Apply the first law to calculate the change in internal energy.
This problem can be simply answered using the first law. This tells us:

Change in internal energy = Heat flow – Work done.

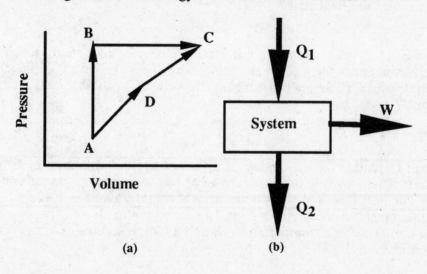

(a)　　　　　　(b)

FIGURE SG 11.1: a). Pressure versus volume diagram for the fluid.
b). Heat and work flow for the system under consideration.

Thus for the path $A \Rightarrow B \Rightarrow C$,

$$\text{Change in internal energy} = +500 \text{ J} - 200 \text{ J} = 300 \text{ J}.$$

Since the internal energy of a system does not depend upon how it got to its present state, we know for the path $A \Rightarrow D \Rightarrow C$ that the change in internal energy is the same as previously calculated. The individual values for heat flow and work done may differ, but their difference will have the same value.

Before leaving this problem, let us examine what is happening in the two situations when we also consider the second law. The second law tells us that the system does not just absorb heat and do work; it must also eject heat at a lower temperature. This is shown in Figure SG 11.1b. Some heat, Q_1, flows into the system and some heat, Q_2, is ejected from the system while work, W, is done. The values of Q_1, Q_2, and W that we have will depend upon the particular path taken from A to B; that is, if we let

$$B \text{ denote the path } A \Rightarrow B \Rightarrow C$$

and

$$D \text{ denote the path } A \Rightarrow D \Rightarrow C,$$

then, in general

$$Q_{1B} \neq Q_{1D}$$
$$Q_{2B} \neq Q_{2D}$$
$$W_B \neq W_D$$

but

$$\Delta U = Q_{1B} - Q_{2B} - W_B = Q_{1D} - Q_{2D} - W_D.$$

That is, the net change in the internal energy of the system, ΔU, is independent of the path used to reach the final state.

PROBLEM SG 11.2: An inventor claims to have developed a steam turbine which is 74% efficient. Steam enters the turbine at a temperature of 300°C and is exhausted as water at a temperature of 30°C. Is such a turbine theoretically possible and is the claimed efficiency correct?

SOLUTION:

> **Procedure:** Calculate the efficiency of a Carnot cycle and compare with the given 74%. Carnot efficiency is less so analyze what factors effecting the turbine have been neglected. The heat flow and work done are shown in Figure SG 11.2.

The maximum thermal efficiency is that of a Carnot engine. If a Carnot engine operates between a hot reservoir at temperature T_H and a cool reservoir of temperature T_C, its efficiency is given by:

$$\text{thermal efficiency}_{\text{maximum}} = 1 - \frac{T_C}{T_H}$$

$$= 1 - \frac{(273+30)\text{K}}{(273+300)\text{K}} = 0.47.$$

Since 47% is less than 74%, we know the 74% efficiency claim is wrong.

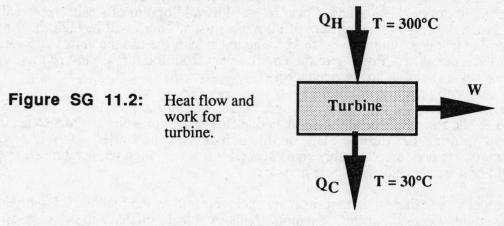

Figure SG 11.2: Heat flow and
work for
turbine.

What might the inventor have done incorrectly to get his 74% figure?

He probably used Equation 11.3 of the text incorrectly. This equation gives

$$\text{thermal efficiency} = \frac{Q_H - Q_C}{Q_H}.$$

He probably identified $Q_H - Q_C$ as the heat gained when 1 kg of steam at
300°C is changed to 1 kg of water at 30°C and computed Q_H by adding the
heat that could be extracted by cooling 30°C water to 0 K ice.

Using tables 10.3 and 10.4 of the text we have:

Heat which is given off by cooling 1 kg of 300°C steam to 30°C water

cool 300°C steam to 100°C	$\Rightarrow 0.48 \times (300-100)$ kcal $\Rightarrow$	96 kcal
convert steam to 100°C water		540 kcal
cool water to 30°C	$\Rightarrow 1.00 \times (100-30)$ kcal	70 kcal
		706 kcal

The heat which can be extracted from 30°C water is:

cool water to 0°C	$1.00 \times (30 - 0)$ kcal	30 kcal
freeze water		80 kcal
cool water to –273.16°C	$0.50 \times (0 - -273.16)$ kcal	137 kcal
		247 kcal

He then erroneously identified

$$Q_H - Q_C = 706 \text{ kcal} \qquad \text{and} \qquad Q_H = 951 \text{ kcal.}$$

and calculated

$$\text{thermal efficiency} = \frac{W}{Q_H} = \frac{706 \text{ kcal}}{951 \text{ kcal}} = 0.74.$$

There are several mistakes he has made, among them are:

1. The heat into and out of the turbine cannot be calculated as he has done. The heat flow will not be the amount of heat released by a kilogram of steam converted to water multiplied by the number of kilograms used. There will be additional heat flowing from the hot reservoir to keep the hot side of the turbine at its temperature and additional heat flowing to the environment from the turbine to keep its outer jacket at the surrounding temperature.

2. No mention is made of possible changes in the volume and pressure of the working fluid. The steam will be at greater than one atmosphere pressure in order to get it to flow through the turbine. The implication is that the outlet water is at one atmosphere pressure. Probably some heat flow from the high temperature reservoir will be needed to equalize these pressures.

The proper way to find Q_H is to measure the heat released by the fuel needed to produce the steam that keeps the turbine running. Q_C is extremely difficult to calculate since there are many energy losses in practice. Experience permits an engineer to make a good estimate of thermal efficiency before a turbine is built.

PROBLEM SG 11.3: A refrigerator maintains inside temperature of 5.0°C when the exterior temperature is 25°C. It maintains the same interior temperature when the room temperature is 30°C. How big is the percentage change in the coefficient of performance for the two cases?

SOLUTION:

Procedure: Calculate the coefficient of performance for the two cases using text equation 11.5.

As is shown in Figure SG 11.3a and Figure SG11.3b, a refrigerator extracts heat from a low temperature reservoir and deposits it into a high temperature reservoir. The second law of thermodynamics tells us that this is only possible if work is done on the refrigerator.

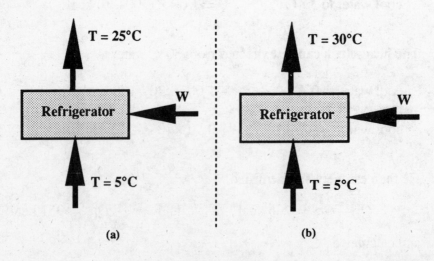

(a)   (b)

Figure SG 11.3   a). Refrigerator operating between 5°C and 25°C.
b). Refrigerator operating between 5°C and 30°C.

The coefficient of performance of a refrigerator is given by Equation 11.5 of the text as:

$$\text{c.p.(refrigerator)} = \frac{T_C}{T_H - T_C}$$

where T_H is the hot temperature and T_C is the cold temperature.

For the first case

$$\text{c.p. } (25°C) = \frac{273 + 5}{(273+25)-(273+5)} = 13.9 = 14,$$

while in the second case

$$\text{c.p. } (30°C) = \frac{273 + 5}{(273+30)-(273+5)} = 11.12 = 11.$$

The percent change in coefficient of performance from the first to the second case is

$$100 \times \frac{14 - 11}{14} = 21.14 \% = 21 \%.$$

Assuming the cost of operation is proportional to the c.p., we see that it will cost nearly 1/5 more to operate the refrigerator at 30°C than at 25°C.

PROBLEM SG 11.4: Problem 10.20 of the text expresses energy in "jelly donut" units of 1 "jelly donut" = 1 MJ. Assuming that jelly donuts cost $1 each, is it worth the effort to pick up a penny? That is, compare the amount of energy expended by picking up a penny ($0.01) with the amount of energy that could be purchased by that penny spent on jelly donuts? Hint: To do this problem you must make some reasonable assumptions about the process of picking up the penny and about your mass. Neglect the mass of the penny and consider only the work you must do to raise and lower your center of mass. Assume a weight of 150 lbs and a squat involves moving your center of mass 0.50 meters up and down.

SOLUTION:

> **Procedure:** Make a simple model of the process of bending over to pick up the penny. Calculate the energy expended in picking up one penny.

The simplest process of picking up a penny is shown in Figure SG 11.4. We will assume the picking up process is done by a squat. A squat moves the center of mass straight up and down. We will make the approximation that it takes no work to smoothly squat down and assume that all the work is done in standing back up. If d is the distance the center of mass moves in the squat and m is your mass, then to to pick up a penny we have

$$\text{work done} = m\, g\, h,$$

where g is the acceleration of gravity.

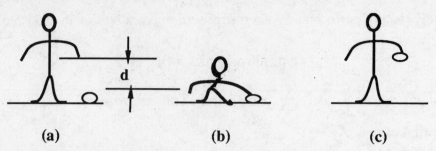

Figure SG 11.4: a)-c). The process of picking up a penny.

STEP 1:

Determine the work done in joules when picking up the penny. Assuming a weight of 150 pounds and a squat distance of about 0.50 m,

$$\text{work done} \;=\; \frac{150 \text{ lbs (weight)}}{32 \text{ ft/s}^2} \times \frac{0.454 \text{ kg}}{\text{lb (mass)}} \times 9.8 \text{ m/s}^2 \times \frac{1}{2}\,\text{m}$$

$$= 10.4278 \text{ J} \;=\; 10 \text{ J}.$$

STEP 2:

Determine the energy in joules which could be purchased by the penny spent on jelly donuts.

$$1 \text{ penny} \;\Rightarrow\; \$0.01 \times \frac{1 \text{ "jelly donut"}}{\$1} \times \frac{1 \text{ MJ}}{1 \text{ "jelly donut"}} \times \frac{10^6 \text{ J}}{\text{MJ}}$$

$$\Rightarrow\; 1 \times 10^4 \text{ J}$$

STEP 3:

Comparing the results we see a gain of 1000:1; therefore, it is "worth" the energy expended to pick up the penny.

Or alternatively, argue that since

$$\frac{1 \text{ MJ}}{10 \text{ J}} \;=\; 10^5,$$

you could pick up 100,000 pennies or $1,000 in pennies on the energy of 1 "jelly donut". This is a 1000% return on your effort.

PROBLEM SG 11.5: Determine the change in entropy that occurs when 30 grams of cream at 5°C is stirred into 400 grams of coffee at 85°C. Assume the specific heat of both the cream and the coffee is essentially the same as that of water and that the stirring process introduces 1,000 J of energy.

SOLUTION:

> **Procedure:** Distribute the stirring energy proportional to the two masses and assume no energy losses to the environment. Use the methods of the previous chapter to find the final energy. Then calculate the entropy by text equation 11.7.

Neglect all heat losses. There is a heat flow into the cream and a heat flow out of the coffee. We will assume the energy of stirring is distributed proportional to the masses of each substance.

The heat flows must be equal, so we can use the methods of chapter 10 to calculate the final temperature. We than find the average temperature for each substance by

$$T_{avg} = \frac{T_{initial} + T_{final}}{2}$$

and use Equation 11.7 of the text to calculate each entropy change.

GIVEN:

Quantity	Symbol	Value
mass of cream	m	30 g
mass of coffee	M	400 g
initial temperature of cream	T_C	5°C = 278 K
initial temperature of coffee	T_H	85°C = 358 K
final temperature of mix	T	unknown
energy of stirring added to cream	ΔQ_1	$\dfrac{30 \times 1000}{400+30}$ J
energy of stirring added to coffee	ΔQ_2	$\dfrac{400 \times 1000}{400+30}$ J
specific heats of both substances	c	1 cal/g–K

Note that knowing the temperature in C° to 2 significant figures is the same as knowing the temperatures to 3 significant figures in K

The heat <u>gained</u> by the cream is

$$\Delta Q_{cream} = \Delta Q_1 + m\,c\,(T - T_C),$$

while the heat <u>lost</u> by the coffee is

$$\Delta Q_{coffee} = -\Delta Q_2 + M\,c\,(T_H - T).$$

Notice the minus sign put in for ΔQ_2 since the energy of stirring is added to the coffee. It is probably most convenient to use calories, grams and Kelvin degrees as our units, since that makes c unity and saves converting temperatures at the end. Expressing ΔQ_1 and ΔQ_2 in cal by the conversion 1 J = 0.2388 cal, we get

$$\Delta Q_{cream} = 16.7 \text{ cal} + 30 \times (T - 278 \text{ K}) \frac{cal}{K}$$

$$\Delta Q_{coffee} = -222.1 \text{ cal} + 400 \times (358 \text{ K} - T) \frac{cal}{K}.$$

Equating these we get

$$(M+m)cT = c(M\,T_H + m\,T_C) - \Delta Q_1 - \Delta Q_2$$

or

$$T = \frac{M\,T_H + m\,T_C}{M + m} - \frac{\Delta Q_1 + \Delta Q_2}{c(M + m)}$$

$$= (352.4 - 0.5)\,K = 352\,K.$$

Thus the final temperature is $352\,K = 79°C$.

The second term is the effect of stirring which is negligible for this example. Remember that K to 3 significant figures is $C°$ to two figures.

The average temperatures are now

$$T_{\text{avg-cream}} = 316\,K$$

and

$$T_{\text{avg-coffee}} = 356\,K.$$

Before we can calculate the entropy change, we must get the (equal) heat flows for each substance. Using the final temperature we get

$$\Delta Q_{\text{cream}} = (16.7 + 2220)\,\text{cal} = 2.2 \times 10^3\,\text{cal},$$

and

$$\Delta Q_{\text{coffee}} = (-222.1 + 2400)\,\text{cal} = 2.2 \times 10^3\,\text{cal}.$$

$$= \Delta Q_{\text{cream}}.$$

These two are equal as expected, since ΔQ_{cream} is <u>heat added </u>to the cream and ΔQ_{coffee} is the <u>heat lost</u> by the coffee.

Before we calculate the entropy change we must be careful and remember that ΔQ_{cream} is energy added and ΔQ_{coffee} is energy lost, so the total entropy change, ΔS, is

$$\Delta S = \frac{\Delta Q}{T}$$

$$= \frac{2.2 \times 10^3\,\text{cal}}{316\,K} - \frac{2.2 \times 10^3\,\text{cal}}{356} = (6.96 - 6.18)\frac{\text{cal}}{K} = 0.78\frac{\text{cal}}{K}.$$

The entropy change is positive as required.

Chapter 12
The Kinetic Theory of Gases

Important Terms

Fill in the blanks with the appropriate word or words.

1. The fraction of molecules in a gas found as a function of temperature and velocity is given by the _____.

2. The relation between the volume and the temperature for an ideal gas at constant pressure is called _____ Law.

3. The number of molecules in a mole of a substance is called _____ number.

4. _____ equation is a simple modification of the ideal gas law that considers the size of molecules.

5. The model of a gas in which the gas is treated as made up of many particles is the _____ .

6. The five principle assumptions of the kinetic theory model of gases define what is called a(n) _____ gas.

7. _____ Law states "The pressure exerted by a gas is inversely proportional to the volume of space in which it is enclosed".

8. The ratio of the heat added per unit volume to the temperature change for a gas at constant pressure is called _____ .

9. The constant appearing in the ideal gas law is called _____ constant.

10. Any equation linking pressure, temperature, and volume is called an equation _____ .

11. The gas constant per molecule is called _____ constant.

12. The relation between the pressure, temperature, number of moles, and the volume of an ideal gas is called the _____ .

13. The _____ is the amount of a substance of a system which contains as many elementary entities as there are atoms in 0.012 kg of carbon 12.

14. The ratio of the heat added per unit volume to the temperature change for a gas at constant volume is called _____ .

Write the definitions of the following:

15. Boyle's Law (12.2):

16. Charles' and Gay-Lussac's law (12.3):

17. Maxwell-Boltzmann distribution (12.9):

18. mole (12.4):

19. ideal gas law (12.4):

20. universal gas constant (12.4):

21. ideal gas (12.5):

22. Boltzmann constant (12.6):

23. molar specific heat at constant pressure (12.7):

24. molar specific heat at constant volume (12.7):

25. equation of state (12.4):

26. Avogadro's number (12.4):

27. van der Waals Equation (12.8):

Answers to 1–14, Important Terms

1. Maxwell-Boltzmann distribution; 2. Charles' and Gay-Lussac's; 3. Avogadro's; 4. van der Waals equation; 5. kinetic theory model; 6. ideal; 7. Boyle's; 8. specific heat at constant pressure; 9. universal gas; 10. of state; 11. Boltzmann; 12. ideal gas law; 13. mole; 14. specific heat at constant volume

Sample Solutions

PROBLEM SG 12.1: What is the total downward atmospheric force on the top of a 2.0 m by 3.0 m table top? Why doesn't this force break the table or its legs?

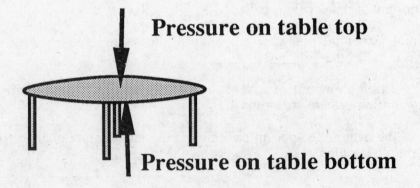

Pressure on table top

Pressure on table bottom

Figure SG 12.1: Table with upward and downward atmospheric forces shown.

SOLUTION:

> **Procedure:** Force equals pressure times area. If there is no motion all forces must add to zero.

Since force is pressure times area, the total downward force on the table top, F, is given by:

$$F = p \times A.$$

Atmospheric pressure is 1.01×10^5 Pa, so the force is:

$$F = p \times A = 1.01 \times 10^5 \text{ Pa} \times (2.0 \text{ m} \times 3.0 \text{ m})$$

$$= 1.01 \times 10^5 \, \frac{\text{N}}{\text{m}^2} \times (2.0 \text{ m} \times 3.0 \text{ m}) = 6.1 \times 10^5 \text{ N}.$$

This downward force does not break the table since it is balanced by an equal upward atmospheric force on the lower side of the table as shown in Figure SG 12.1.

PROBLEM SG 12.2: An upright glass cylinder 75 cm tall and 3.0 cm in diameter is fitted with a light piston which is free to slide. The cylinder is closed at its lower end. First the piston is placed on the cylinder. Initially the piston sinks to 73 cm from the bottom of the cylinder. Water is poured into the cup-like cavity formed by the top of the piston and the cylinder until the cavity is full. At what fraction of the total height of the cylinder will the piston be? Assume the lower portion of the cylinder contains an ideal gas at constant temperature.

SOLUTION:

> **Procedure:** Since the temperature is assumed constant, this should be an application of Boyle's law. That is, pressure times volume is constant. The gravitational force on the water provides an added pressure on the gas trapped below the piston.

The physical arrangement before the piston is added is shown in Figure SG 12.2a, before the water is added is shown in Figure SG 12.2b, and after the water is added is shown in Figure SG 12.2c.

Since the area of the piston is constant, we can expect the diameter of the cylinder not to appear in the answer.

We define the following variables [Note that the area is $\pi \, (\text{diameter}/2)^2$]:

GIVEN:

Quantity	Symbol	Value
area across cylinder	A	7.0686 cm^2
length of cylinder	L	75 cm = 0.75 m
initial height of piston above bottom of cylinder	h_0	73 cm = 0.73 m
final height of piston above bottom of cylinder	h	unknown
density of water	ρ	1000 kg/m^3
pressure in gas after piston is added	P_i	unknown
atmospheric pressure	p_a	1.01325×10^5 Pa

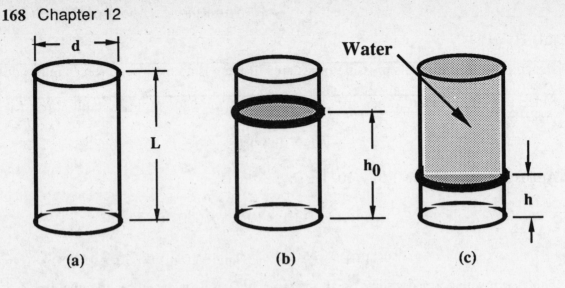

Figure SG 12.2:   a). Cylinder before piston is added.
b). Piston in cylinder before the water is added.
c). Piston and cylinder after water is added.

We assume that the pressure of the ideal gas in the bottom of the cylinder was initially at one atmosphere.

FIRST STEP
The gas was compressed by the weight of the piston until the gas pressure was able to support the piston. This occurred when the piston was a height h_0 above the bottom. The law which covers this behavior at constant temperature is Boyle's Law, which states

$$PV = \text{constant}, \qquad \text{(text Equation 12.1)}$$

where P is the gas pressure and V is the gas volume.

SECOND STEP
We now add water. The weight of the water will compress the gas in the piston. The compressed gas will have a higher pressure. When the gas is compressed enough, this pressure will balance force provided by the added water. Of course, more water will be needed than the amount that could originally fit into the "cup" since the piston is now lower in the cylinder. Finally, the piston will be in balance between the opposing forces of the trapped gas and the piston and water.

Before the water is added, Boyle's law gives

$$P_i V = P_i \times h_0 \times A = \text{constant}.$$

The weight of the added water is $(L - h) \times g \times \rho \times A$, so the pressure added by the water is

$$\frac{(L - h) \, g \, \rho \, A}{A} = \rho \, g \, (L - h).$$

Hence,

$$(P_i + \rho\, g\, [L - h]\,) \times h \times A \;=\; \text{constant} \;=\; P_i \times h_0 \times A.$$

At first inspection we seem to have one equation with two unknowns (h and P_i). This apparent difficulty can be resolved by realizing that before the piston was added, the pressure of the gas in the cylinder was atmospheric.

Let us use P_p to represent the pressure added to the gas by the piston. That is,

$$P_p \;=\; \frac{\text{weight of piston}}{\text{area of piston}},$$

then we have the following three equations with the same constant:

Before the piston is added,

$$P_a \times (L \times A) \;=\; \text{constant}. \tag{I}$$

where the choice of length $= L$ is made since this is the gas trapped by the piston.

After the piston is added,

$$(P_a + P_p) \times (h_0 \times A) \;=\; \text{constant}. \tag{II}$$

After the water is added,

$$(P_a + P_p + \rho g[L - h]) \times (h \times A) \;=\; \text{constant}. \tag{III}$$

Equation III follows from the weight of the water being $\rho g(L - h)A$ as is shown in Figure SG 12.2.a. Hence the pressure due to the water is $\rho g(L - h)$.

The equations I and II give,

$$P_a + P_p \;=\; P_a \frac{L}{h_0}.$$

While equations I and III give,

$$[P_a + P_p + \rho g(L - h)] \times h \;=\; P_a L.$$

We can thus eliminate the effect of the piston and get

$$\left(\frac{P_a L}{h_0} + \rho g\,[L - h] \right) h \;=\; P_a L.$$

The solution to this quadratic equation is:

$$h = \frac{\left(\frac{P_aL}{h_0} + \rho g L\right) \pm \sqrt{\left(\frac{P_aL}{h_0} + \rho g L\right)^2 - 4\rho g L P_a}}{2\rho g}$$

$$= Z \pm \sqrt{Z^2 - \frac{LP_a}{\rho g}}$$

where

$$Z = \frac{\left(\frac{P_aL}{h_0} + \rho g L\right)}{2\rho g} = \frac{P_aL}{2\rho g h_0} + \frac{L}{2}.$$

Now

$$\frac{P_aL}{2\rho g h_0} = 5.3 \frac{N \, s^2}{kg} = 5.3 \, m,$$

$$\frac{L}{2} = 0.37 \, m,$$

and

$$\frac{LP_a}{\rho g} = 7.6 \, m^2.$$

Thus

$$h = 5.7 \, m \pm \sqrt{(5.7 \, m)^2 - 7.6 \, m^2} = 11 \, m \text{ or } 0.71 \, m.$$

The 11 m answer is clearly not correct, so we conclude that when the water is poured into the top of the cylinder, the piston sinks to 70 cm from the bottom of the cylinder.

PROBLEM SG 12.3: A column of dry air was sealed off from the atmosphere by closing one end of a glass tube and placing a drop of mercury in the tube 0.40 m from the end of the tube. This was done at a temperature of 30°C. The tube was then placed in a freezer whose temperature is unknown. After the tube reached the temperature of the freezer, the mercury slug was found to be 0.35 m from the closed end. What is the temperature of the freezer?

SOLUTION:

> **Procedure:** Since this is a constant pressure (isobaric) experiment, use the law of Charles-Gay-Lussac.

Figure SG 12.3 shows the tube at 30°C in 12.3a and at the unknown temperature in 12.3b. Since in both cases, one end of the tube is open, the external pressure is atmospheric. We must assume the atmospheric pressure outside and inside the freezer is the same. The mercury slug is free to move, so the pressure of the gas trapped inside the tube is constant as the tube and gas cool.

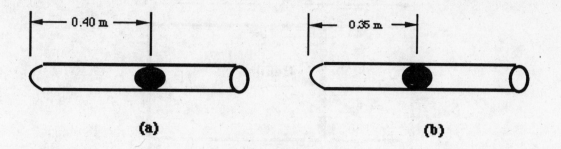

(a)   **(b)**

Figure SG 12.3:   a). Tube with mercury slug at 30°C.
b). Tube with mercury slug at unknown temperature.

Since the experiment is carried out at constant pressure, the applicable relation is the law of Charles-Gay-Lussac as given in the text Equation 12.3. Thus

$$\frac{V}{T} = \text{constant},$$

where V is the volume of the trapped gas and T is its temperature.

If we neglect the small decrease in the internal diameter of the glass tube as it cools, the volume of the gas is proportional to the length of the gas column, so

$$\frac{\text{length at } 30°C}{273 + 30} = \frac{\text{length at } T°C}{273 + T}.$$

Hence

$$T = \left((273 + 30)\frac{\text{length at } T°C}{\text{length at } 30°C} - 273\right)°C$$

$$= \left(303\frac{0.35 \text{ m}}{0.40 \text{ m}} - 273\right)°C$$

$$= [265.125 - 273]°C = -7.875°C = -8.0°C.$$

The temperature of the freezer is −8.0°C.

PROBLEM SG 12.4: A cylinder closed at both ends has a piston in between which is free to slide. At 20°C the piston is exactly one third of the way from the left end when hydrogen gas is in the left side and helium is in the right side. What is the ratio of the number of hydrogen molecules to the number of helium molecules?

SOLUTION:

> **Procedure:** Since we have different gases on each side of the piston, we will need to use the ideal gas law since the number of moles enters into that law's formulation. Equate temperatures and pressures to find the relative number of moles.

By the statement of the problem, both gases have the same temperature and the same pressure. They have different volumes as shown in Figure SG 12.4.

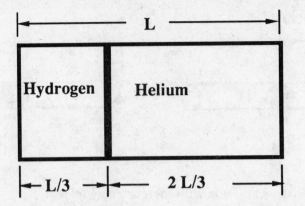

Figure SG 12.4: Cylinder containing hydrogen and helium separated by a piston.

We will assume that hydrogen and helium both obey the ideal gas law, that is Equation 12.4 of the text,

$$PV = nRT,$$

where P is the pressure, V the volume, n the number of moles, T the temperature, and R is a constant. Thus the number of moles we have for an ideal gas is

$$n = \frac{PV}{RT}.$$

If we let the subscript hy denote the hydrogen and the subscript he denote the helium, we have

$$n_{hy} = \frac{P_{hy}V_{hy}}{RT_{hy}}$$

and

$$n_{he} = \frac{P_{he}V_{he}}{RT_{he}}.$$

So

$$\frac{n_{hy}}{n_{he}} = \frac{V_{hy}}{V_{he}} = \frac{1}{2}.$$

The mole is a count of molecules, so the ratio of moles is the same as the ratio of the number of molecules; that is, there are twice as many helium molecules as hydrogen molecules. Note however that a hydrogen molecule has two atoms and helium is monoatomic, so in this case there are the same number of atoms.

PROBLEM SG 12.5: What is the average kinetic energy of a carbon dioxide molecule at 300 K? Assume carbon dioxide at this temperature is an ideal gas. What is the root mean square average speed of a molecule?

SOLUTION:

> **Procedure:** Use the results of the kinetic theory of gases, in particular text Equation 12.8.

The kinetic theory of gases gives the relation between average kinetic energy (KE_{avg}) and the temperature (T) in text Equation 12.8, then

$$KE_{avg} = \frac{3}{2} k T.$$

Carbon dioxide has a molecular mass of 44 g/mol (see text Table 12.1). If we denote this by M_c, then the actual mass of a single molecule would be

$$m_c = M_c/N_A \text{ (expressed in grams)}$$

where N_A is Avogadro's number.

In terms of the root mean square speed, v_{rms}, the average kinetic energy is

$$KE_{avg} = \frac{1}{2} m_c v_{rms}^2,$$

or

$$v_{rms} = \sqrt{\frac{2\, KE_{avg}}{m_c}}.$$

To summarize,

GIVEN:

Quantity	Symbol	Value
Temperature	T	300 K
mass of a CO_2 molecule	m_c	$44/N_A$ = 7.31×10^{-23} g 7.31×10^{-26} kg
Boltzmann constant	k	1.3807×10^{-23} JK^{-1}molecule^{-1}

Then

$$KE_{avg} = \frac{3}{2} k T$$

$$= \frac{3}{2} \times 1.3807 \times 10^{-23} \text{ JK}^{-1}\text{molecule}^{-1} \times 300 \text{ K}$$

$$= 6.21315 \times 10^{-21} \text{ J molecule}^{-1} = 6.2 \times 10^{-21} \text{ J molecule}^{-1}$$

and

$$v_{rms} = \sqrt{\frac{2\, KE_{avg}}{m_c}}$$

$$= \sqrt{\frac{2 \times 6.21315 \times 10^{-21} \text{ J}}{7.31 \times 10^{-26} \text{ kg}}}$$

$$= 4.123 \times 10^2 \text{ km/s} = 4.1 \times 10^2 \text{ km/s}.$$

The average kinetic energy per molecule is 6.2×10^{-21} J which corresponds to an rms speed of 4.1×10^2 km/s.

PROBLEM SG 12.6: What is the ratio of the temperature change in one mole of an ideal gas after 10 Joules are added at constant pressure to the temperature change if the 10 Joules are added at constant volume? What is the temperature change in each case?

SOLUTION:

Procedure: Use the definitions of the two specific heats and the numerical values from the text to find the ratio of the temperature changes.

The specific heat at constant volume, C_V, is given as

$$C_V = 12.47 \frac{J}{mol\ K}$$

and the specific heat at constant pressure, C_P, is given as

$$C_P = 20.78 \frac{J}{mol\ K}.$$

The temperature change, ΔT, is given in terms of the energy added, ΔQ, and the number of moles, n, by

$$\Delta Q_V = nC_V\Delta T_V,$$

for a constant volume process or by

$$\Delta Q_P = nC_P\Delta T_P,$$

for a constant pressure process. The subscript indicates the process used to add the energy.

The ratio of temperature changes is then

$$\frac{\Delta T_P}{\Delta T_V} = \frac{\dfrac{\Delta Q_P}{nC_P}}{\dfrac{\Delta Q_V}{nC_V}} = \frac{C_V}{C_P} = 0.6001.$$

Note that the result is independent of the number of moles and the amount of heat added, as long as the same amount of heat is added in each process.

The temperature change depends upon the heat added and the number of moles. For our problem, $n = 1$ and $\Delta Q = 10$ J, so

$$\Delta T_P = 0.48\ K \quad \text{and} \quad \Delta T_V = 0.80\ K.$$

PROBLEM SG 12.7: Calculate the height at which the atmospheric pressure is one third (0.333) the sea level pressure P_O for air with average molecular mass of 28.9 and a uniform temperature of 295 K.

SOLUTION:

Procedure: Use the exponential law for the pressure with height as given in text equation 12.14.

The dependence of pressure, P, upon height is given in the text by Equation 12.14 as

$$P(z) = P_O\, e^{-mgz/kT},$$

where m is the mass of a molecule, g is the acceleration of gravity, z is the height, k is Boltzmann's constant and T is the temperature.

GIVEN:

Quantity	Symbol	Value
height	z	unknown
sea level pressure	P_O	
pressure	$P(z)$	$0.333\, P_O$
temperature	T	295 K
acceleration of gravity	g	9.80 m/s^2
Boltzmann's constant	k	1.3807×10^{-23} JK^{-1}molecule^{-1}
molecular mass	M	28.9
mass of molecule	$m = M/N_A$	4.80×10^{-23} g $= 4.80 \times 10^{-26}$ kg.

Calculate

$$\frac{1}{z_O} = \frac{mg}{kT} = \frac{4.80 \times 10^{-26}\text{ kg} \times 9.80\text{ m/s}^2}{1.3807 \times 10^{-23}\text{ JK}^{-1}\text{molecule}^{-1} \times 295\text{ K}}$$

$$= 1.1547 \times 10^{-4}\text{ m}^{-1} = 1.2 \times 10^{-4}\text{ m}^{-1}.$$

Then

$$P(z) = 0.333\, P_O = P_O\, e^{-z/z_O},$$

or

$$0.333 = e^{-z/z_O}.$$

The inverse of the operation e^x is denoted by ln. That is, if

$$y = e^x \qquad \text{then} \qquad ln(y) = x.$$

For example if $x = 2$, then $y = e^x = 7.3890561$, so $ln(7.3890561) = 2$.

Most calculators have a ln key (this key is not the same as the log key). Use your calculator to find $ln(3) = 1.0986123$ and $ln(12) = 2.4849066$. Your calculator may round off these numbers with fewer figures.

Returning to our problem, use your calculator to find

$$ln(0.333) = -1.0996128,$$

so

$$-1.0996128 = -z/z_0 = -z \times 1.2 \times 10^{-4} \text{ m}^{-1},$$

or

$$z = 9.5 \times 10^3 \text{ m}.$$

The atmospheric pressure is 1/3 that of sea level at 9.5 km with the molecular mass and temperature we assumed.

Chapter 13
Periodic Motion

Important Terms

Fill in the blanks with the appropriate word or words.

1. A physical phenomenon which repeats itself regularly is termed _____.

2. Another name for the natural frequency of a system is the _____ frequency. This term is used because the amplitude of the system increases rapidly at that frequency.

3. The angle by which a harmonic function is offset at the time chosen as zero is called the _____ angle.

4. When two waves are in phase the amplitude builds up. This is called _____ .

5. The force that a displaced stretch string exerts to return to its undistorted position is called a(n) _____ force.

6. If an oscillator also has frictional forces, we refer to these forces as _____ forces.

7. Motion whose acceleration is proportional to displacement is called _____ motion.

8. The period of a freely oscillating system is called the _____ period.

9. The _____ is the frequency of a freely oscillating system.

10. The proportionality constant in Hooke's Law is called a(n) _____ .

11. A(n) _____ consists of a mass swinging from the end of a light string of constant length.

12. The maximum value of displacement of a wave is called the _____ of the wave.

13. Any object whose motion obeys Hooke's Law is called a simple _____ oscillator.

Write the definitions of the following:

14. amplitude (13.2):

15. resonance (13.11):

16. spring constant (13.1):

17. restoring force (13.1):

18. resonant frequency (13.7):

19. periodic (13.0):

20. damping (13.6):

21. natural frequency (13.4):

22. phase angle (13.2):

23. simple harmonic motion (13.2):

24. natural period (13.4):

25. simple pendulum (13.5):

26. simple harmonic oscillator (13.2):

Answers to 1–13, Important Terms

1. periodic; 2. resonant; 3. phase; 4. resonance; 5. restoring; 6. damping; 7. simple harmonic; 8. natural; 9. natural frequency; 10. spring constant; 11. period; 12. amplitude; 13. harmonic

Sample Solutions

PROBLEM SG 13.1: A mass of 0.140 kg is hung on the end of a spring as shown in Figure SG 13.1a. At the new equilibrium point the spring is stretched 12 .0 cm from its initial length. The mass is now pulled down a further 15.0 cm and released at time $t = 0$. Answer the following:

    a). What is the spring constant?

    b). What is the period of oscillations?

    c). What is the general form of the motion after release.

    d). What are the displacement, velocity and acceleration at

$$t = t_1 = 0.100 \text{ s}, t = t_2 = 0.300 \text{ s}, t = t_3 = 0.520 \text{ s}?$$

SOLUTION:

> **Procedure:** This problem involves calculating the spring constant using Hooke's law; then using that to find the period and hence the displacement.

In order to express our ideas, we will make the drawing shown in Figure SG 13.1a. The following variables will be used:

Quantity	Symbol	Value
spring constant	k	unknown
mass added	m	0.140 kg
change in spring length	d	12.0 cm
further initial displacement	x_0	15.0 cm
later displacement at time t	$x(t)$	unknown
velocity at time t	$v(t)$	unknown
acceleration at time t	$a(t)$	unknown
period of oscillation	τ	unknown
acceleration of gravity	g	9.80 m/s^2

When we hang the mass on the spring, the spring stretches until the Hooke's Law force plus the gravitational force is zero. That is

$$F = -k\,d, \qquad \text{(Hooke's Law)} \qquad \text{(text Equation 13.1)}$$

and

$$F + mg = 0 \qquad \text{(condition for equilibrium)}$$

Thus

$$k = \frac{mg}{d}$$

$$= 0.140 \text{ kg} \times \frac{9.80 \text{ m/s}^2}{0.12 \text{ m}} = 11.4333 \text{ N/m.} = 11.4 \text{ N/m.}$$

Considering the significant figures, the spring constant $k = 11.4$ N/m.

Since the gravitational force is constant, this initial stretching of the spring will always take care of the gravitational force. So the subsequent motion of the mass is not affected by the gravitational force, provided we measure the displacement from the new equilibrium position. This is the choice we have made for $x(t)$ in Figure SG 31.1.

The period of oscillation is given by text Equation 13.8 as

$$\tau = 2\pi \sqrt{\frac{m}{k}} = 2\pi \sqrt{\frac{m}{mg/d}}$$

$$= 2\pi \sqrt{\frac{d}{g}} = 0.69528 \text{ s,}$$

or, considering significant figures, $\tau = 0.695$ s.

Note that for this problem, as in the case of the simple pendulum, the motion is independent of the mass when measured relative to the equilibrium point.

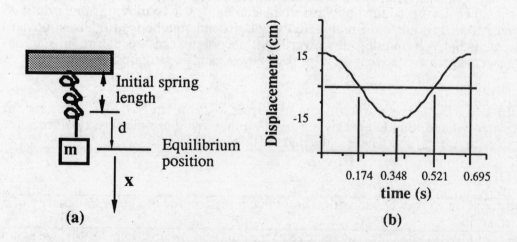

(a)

(b)

Figure SG 13.1   a). Mass of 1.0 kg hanging on a spring. b). Time displacement of the mass following the initial displacement.

The general displacement, velocity and acceleration will be given by equations 13.3 to 13.6 of the text:

$$x(t) = x_0 \cos\left(2\pi \frac{t}{\tau}\right)$$

$$= 0.15 \cos(9.04 \times t) \times m$$

$$v(t) = -\sqrt{\frac{k}{m}}\, x_0 \sin\left(2\pi \frac{t}{\tau}\right)$$

$$= -1.35 \sin(9.04 \times t) \times m/s$$

$$a(t) = -\frac{k}{m}\, x_0 \cos\left(2\pi \frac{t}{\tau}\right)$$

$$= -12.2 \cos(9.04 \times t) \times m/s^2 \ .$$

Figure SG 13.1b shows the general displacement curve and the time values of the critical points.

We can now answer the last question.

TIME	POSITION	VELOCITY	ACCELERATION
0.100 s	0.0928 m	−1.06 m/s	−7.54 m/s^2
0.300 s	−0.136 m	−0.562 m/s	1.23 m/s^2
0.520 s	−0.00174 m	1.35 m/s	0.142 m/s^2

Since we have chosen x positive downwards, positive velocity and positive acceleration are also downwards. So when 0.100 second has passed, we are still below equilibrium but moving upwards with upwards acceleration. After 0.300 seconds the mass has almost reached its maximum upward displacement so it is moving upwards with downward acceleration. And finally, after 0.520 seconds the body still has an upward displacement but it is moving downward, with a downwards acceleration. Each of these answers agrees with what we expect when we examine Figure SG 13.1b.

PROBLEM SG 13.2: For the harmonic oscillator in Problem SG 13.1, what are the maximum potential and kinetic energies of oscillation? What are the first two times the maximum of each occurs, and where is the mass at that time?

SOLUTION:

Procedure: Use energy conservation and the expression for the potential energy of a stretched spring.

We are asked for potential energy relative to the equilibrium position. That is, we again do not need to include the gravitational potential energy of the mass provided the position is measured relative to the equilibrium position. With this in mind, we see that the maximum potential energy, due only to the spring force, occurs first when we start. The maximum potential energy is given by

$$PE_{max} = \frac{1}{2} k \, x_0{}^2.$$

This energy will be a maximum when the spring has maximum extension or maximum compression (since the potential is not changed by $x_0 \to -x_0$). When the potential energy is at maximum, the mass is at rest, so the kinetic energy is zero.

Whenever the mass passes through the equilibrium point the potential energy is zero and hence, since the total energy is conserved, the kinetic energy is a maximum. Thus

$$KE_{max} = PE_{max}.$$

It will take one quarter cycle for the motion to reach the original equilibrium position and that position will be reached every half cycle (see Figure SG 13.1b), so

$$KE_{max} = PE_{max} = \frac{1}{2} \times 11.4 \text{ N/m} \times (0.15 \text{ m})^2 = 0.128 \text{ J}$$

and

Elapsed Time (periods)	Energy Max	Time (seconds)
0.0	PE	0.0
0.25	KE	0.174
0.50	PE	0.348
0.75	KE	0.521

PROBLEM SG 13.3: A 66 pound child (mass = 30 kg) is playing on a swing of negligible mass and of length 3.6 m. The child is being pushed by her father each time she swings and is now swinging so high that her mass reaches a distance of 1.5 m above her starting position. How fast is she moving when she passes through the bottom of her swing? How often must her father push her?

SOLUTION:

Procedure: First use conservation of energy to find the kinetic energy at the bottom of the swing. Calculate the period from the small angle approximation for the simple pendulum.

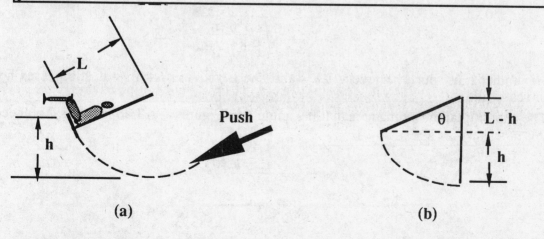

(a) (b)

Figure SG 13.2:   a). Girl on swing.
                            b). Evaluation of the small angle approximation.

We can answer the first question without approximation by using conservation of energy. At any point on her swing her potential energy is given by

$$PE = mgh,$$

where m is her mass, g is the acceleration of gravity and h is her height above the lowest point as shown in Figure SG 13.2a. At the highest point, all the energy is potential energy since her motion is undergoing reversal. This maximum potential energy must be numerically equal to her kinetic energy at the bottom of her swing.

GIVEN:

Quantity	Symbol	Value
mass of girl	m	30 kg
length of swing	L	3.6 m
height of swinging	h	1.5 m
acceleration of gravity	g	9.8 m/s^2
maximum velocity	v_{max}	?

$$KE_{max} = \frac{1}{2} m (v_{max})^2$$

$$= PE_{max} = mgh,$$

so

$$v_{max} = \sqrt{2gh}$$

$$= \sqrt{2 \times 9.8 \text{ m/s}^2 \times 1.5 \text{ m}} = 5.4 \text{ m/s}.$$

She is moving 5.4 m/s at the bottom of her swing.

If we assume that the simple pendulum approximation is close enough, then the period of the swing is given by

$$\tau = 2\pi \sqrt{\frac{L}{g}}$$

$$= 2\pi \times \sqrt{\frac{3.6 \text{ m}}{9.8 \text{ m/s}^2}} = 3.8 \text{ s}.$$

Her father must push her every 3.8 s and she is moving 5.4 m/s as she passes her lowest point.

The approximation we made is that the angle θ in Figure SG 13.2b is small. To check this note that

$$\cos \theta = \frac{L - h}{L}$$

$$= \frac{2.1}{3.6} = 0.58333,$$

which gives

$$\theta = 54°.$$

So θ is not small so the assumption of small swing angle is not very good, but it is the best that can be done at this level of mathematics.

PROBLEM SG 13.4: A mass on the end of a spring is found to have a period of 2.00 s when displaced vertically. When the same mass and spring are placed horizontally on a fairly smooth surface, the amplitude is observed to decrease by 50% after 20 oscillations and that this takes a total time of 41.0 s. What is the damping constant for the second situation? What is the percentage change in the period caused by the damping?

SOLUTION:

> **Procedure:** Assume the motion is described by text Equation 13.12 to find the damping constant. The observed time and number of periods gives us the new period.

The two cases are shown in Figure SG 13.3. Figure SG 13.3a shows the vertical oscillations and Figure SG 13.3b shows the second case with the oscillator horizontal.

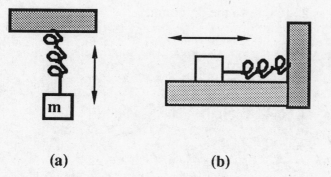

(a)   (b)

Figure SG 13.3:   a). Oscillator is suspended vertically. b). Oscillator is displaced while horizontal on a smooth surface.

Since the surface is smooth and 20 cycles are needed for a 50% decrease in amplitude, the damping force is small, so we can use text Equation (13.12):

$$y = y_0 \, e^{-\gamma t} \cos (2\pi f t).$$

where γ is the damping constant, y is the displacement, t is the time, and f is the frequency of the damped oscillator. Since the damping is small, we expect the f is nearly the natural frequency. However, we can find the damping constant without approximating the frequency, since we are given that the amplitude decreases by 50.0% when t = 41.0 s. That is,

$$\frac{y}{y_0} = 0.500 = e^{-\gamma \times 41.0 \text{ s}}.$$

If we take the natural logarithm of both sides we get

$$\ln(e^{-\gamma \times 41.0\ s}) = -41.0\ s \times \gamma = \ln(0.50) = -0.6931$$

or

$$\gamma = \frac{-0.6931}{-41.0\ s} = 0.0169/s.$$

The damping constant is 0.0169/s.

To find the percentage change in the period, we note that 20 cycles with damping takes 41.0 s. So the damped period is

$$\tau_{damped} = \frac{41.0\ s}{20} = 2.05\ s,$$

while

$$\tau_{undamped} = 2.00\ s.$$

Thus the percentage change is

$$100 \times \frac{\tau_{undamped} - \tau_{damped}}{\tau_{undamped}} = 2.5\%.$$

The period changes $2\frac{1}{2}$ % due to the damping.

Chapter 14
Wave Motion

Important Terms

Fill in the blanks with the appropriate word or words.

1. For a standing wave on a string, the point on the string that vibrates with the greatest amplitude is called a(n) _____ .

2. The _____ frequency of a string is its lowest resonant frequency.

3. Waves in which the propagating medium moves back and forth along the direction of propagation are called _____ waves.

4. A(n) _____ wave pulse is one for which the displacement of the pulse is perpendicular to the direction the pulse travels.

5. If a source is moving through a medium at a speed greater than the wave velocity for the medium, the source produces a single conical wave front known as a(n) _____ .

6. Waves in which the propagating medium moves back and forth perpendicular to the direction of propagation are called _____ waves.

7. When two or more waves combine such as to cancel each other's amplitude, we have _____ interference.

8. For a standing wave on a string, the point on the string that vibrates with the least amplitude is called a(n) _____ .

9. A(n) _____ is a disturbance that transfers energy through a medium without imparting net motion to the medium.

10. If wave fronts are nearly flat, we call the wave a(n) _____ wave.

11. If waves combine in phase so their amplitudes reinforce, we have _____ .

12. A wave generated by simple harmonic motion is called a(n) _____ wave.

13. The _____ frequencies of a string are integer multiples of the fundamental frequency of that string.

14. The _____ of a wave is defined as the power flowing per unit area.

15. At any instant the resultant of a number of waves is the algebraic sum of all the component waves. This is a statement of the _____ .

16. The common unit of sound measurement on the logarithmic scale is the _____ .

17. The compressions (or the rarefactions) produced by a sound wave are referred to as the _____.

18. The _____ is the change in frequency due to relative motion of source and receiver.

19. The separation between successive wave crests or wave fronts is referred to as the _____ of the wave.

20. Relative _____ level is usually measured in decibels.

21. A resonant frequency above the fundamental frequency is called a(n) _____.

22. Two sound waves of nearly the same frequency produce _____.

23. For a(n) _____ wave on a string there exist fixed points on the string where the amplitude is zero and other fixed points where the amplitude is maximum.

24. A(n) _____ wave is a physical effect that travels from one location to another.

Write the definitions of the following:

25. transverse wave pulse (14.1):

26. wave (14.1):

27. harmonic wave (14.2):

28. wavelength (14.2):

29. intensity (14.3):

30. wave front (14.4):

31. plane wave (14.4):

32. shock wave (14.7):

33. harmonic wave (14.2):

34. principle of superposition (14.9):

35. transverse wave (14.4):

36. traveling wave (14.1):

37. constructive interference (14.9):

38. decibel (14.5):

39. fundamental frequency (14.9):

40. standing wave (14.9):

41. nodes (14.9):

42. antinode (14.9):

43. overtone (14.9):

44. sound intensity level (14.5):

45. harmonic frequency (14.9):

46. longitudinal wave (14.4):

47. beats (14.11):

48. Doppler effect (14.6):

Answers to 1–24, Important Terms

1. antinode; 2. fundamental; 3. longitudinal; 4. transverse; 5. shock wave; 6. transverse; 7. destructive; 8. node; 9. wave; 10. plane; 11. constructive interference; 12. harmonic; 13. harmonic; 14. intensity; 15. principle of superposition; 16. decibel; 17. wave fronts; 18. Doppler shift; 19. wavelength; 20. sound intensity; 21. overtone; 22. beats; 23. standing; 24. traveling

Comments on Models

Up to this point the authors of this text have considered the properties of waves in several media — air, water, springs. From these considerations we extract the abstract concept of a wave. The abstract idea of a wave will return in later chapters with electromagnetic waves and quantum matter waves.

As is stated in the text, there are two types of waves: transverse waves and longitudinal waves. Ripples on the surface of a pond are a good example of a transverse wave while a low intensity sound wave provides an example of a longitudinal wave. In both of these examples we restrict our consideration to small disturbances. That is, if you splash the pond hard enough, water will travel horizontally as well as vertically. If you sneeze, there is a net motion of air towards the listener as well as a compression and rarefaction.

In a transverse wave, the material that is carrying the wave moves perpendicular to the direction the wave travels. The surface of a pond moves up and down as a ripple travels across the pond in a horizontal direction. In a sound wave, on the other hand, the air moves back and forth in the same direction as the sound propagates.

Sample Solutions

PROBLEM SG 14.1: A 3000 Hz sound wave passes from air into water. What is the frequency in the water? What is the ratio of the wavelength in air to that of the wave in the water? Assume the temperature is 15°C.

SOLUTION:

> **Procedure:** Use the equation relating wavelength, frequency and velocity along with simple conservation of compressions across a surface.

If we stand at the air/water interface, we see that the frequency is unchanged as the sound wave passes from the air to the water or we would sometimes see a compression of air corresponding to a rarefaction of water or conversely. Thus the frequency in the water is the same as the frequency in the air as shown in Figure SG 14.1.

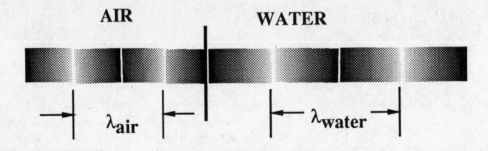

Figure SG 14.1: Sound wave passing from air to water.

This conclusion about the frequencies can not hold for the wavelengths since the frequency, f, wavelength, λ, and the velocity, v, are related by

$$v = \lambda f,$$

or   (text Equation 14.2)

$$\lambda = \frac{v}{f}.$$

We can use text Table 14.1 to get the velocities giving us

$$\frac{\lambda_{air}}{\lambda_{water}} = \frac{\dfrac{v_{air}}{f_{air}}}{\dfrac{v_{water}}{f_{water}}}$$

$$= \frac{v_{air}}{v_{water}} = \frac{340 \text{ m/s}}{1500 \text{ m/s}} = 0.227 .$$

The wavelength in the air is 0.227 times the wavelength in water.

PROBLEM SG 14.2: A train approaching at a speed of 40 m/s sounds as if it has a horn frequency of 1500 Hz. What is its true (rest) frequency? What frequency do you hear as the train recedes? Assume the temperature is 15°C.

SOLUTION:

Procedure: Apply the Doppler effect rule for a moving source.

The two situations are shown in Figure SG 14.2a and 14.2b. Text equations 14.5 apply. If we let f be the true frequency, f' be the frequency we hear, v the relative velocity and v_0 the velocity of sound in air, then as the train approaches

$$f_{app} = \frac{f}{(1 - v/v_0)}$$

and as it recedes

$$f_{rec} = \frac{f}{(1 + v/v_0)}$$

GIVEN:

Quantity	Symbol	Value
frequency heard at approach	f_{app}	1500 Hz
true frequency	f	unknown
frequency heard at recession	f_{rec}	unknown
relative speed	v	40 m/s
speed of sound at 15°C	v_0	340 m/s

The first step is to find f using the first Doppler effect relationship and then use this f to find f'_{rec}.

$$f = f_{app}(1 - v/v_0) = 1500 \text{ Hz} \left(1 - \frac{40 \text{ m/s}}{340 \text{ m/s}} \right),$$

$$= 1.3235294 \times 10^3 \text{ Hz} = 1.3 \times 10^3 \text{ Hz}.$$

The rest frame or true frequency is 1.3×10^3 Hz.

Knowing the true frequency, we can calculate the frequency heard when the train is receding:

$$f_{rec} = \frac{1323.5294}{1.1176471} = 1.184 \times 10^3 \text{ Hz} = 1.2 \times 10^3 \text{ Hz}.$$

We conclude that the horn is emitting a frequency of about 1300 Hz and that we hear a 1200 Hz signal as the train recedes.

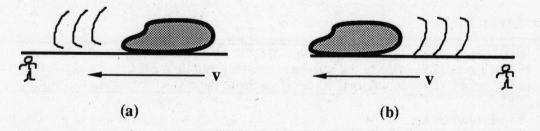

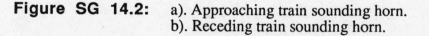

(a)   (b)

Figure SG 14.2:   a). Approaching train sounding horn.
b). Receding train sounding horn.

PROBLEM SG 14.3: A wire is stretched between two supports 1.100 m apart. If the speed of sound in the wire is 2040 m/s, what is the frequency of the fundamental vibration and the first overtone? If two identical such strings are plucked with the first producing the fundamental and the second the first overtone only, what beat frequencies will be heard?

SOLUTION:

> **Procedure:** The frequencies of a string are given in the text by Equation 14.8. Use that to calculate the frequencies and then the beat frequency. Think about the results.

The fundamental frequency of a string is given by

$$f_{\text{fundamental}} = \frac{v}{2L},$$

where v is the sound velocity in the string and L is the length, while the first overtone is given by

$$f_{\text{first overtone}} = \frac{v}{L}.$$

For this string these frequencies are 927.3 Hz and 1.855×10^3 Hz.

The beat frequency is given by the difference in these two frequencies or by

$$f_{\text{beat}} = |f_{\text{fundamental}} - f_{\text{first overtone}}|$$

$$= 927.7 \text{ Hz} = f_{\text{fundamental}},$$

where the last equality follows from the two equations without using the given numerical values. That is, we are caught in significant figures. Since we are subtracting two numbers we loose accuracy and must consider 927.3 and 927.7 the same number.

But this means there is no beat effect since the beat frequency is one of the frequencies with which we began. If we had only one string with the fundamental and any number of overtones excited, no beats will be heard. Beats only are heard when the difference in frequencies is not one of the frequencies present.

PROBLEM SG 14.4: A given sound level is quoted as –2.0 dB above the reference. What would the sound level be if the intensity were increased by a factor of 8.0? What is the original intensity in W/m^2 if the reference intensity is 10^{-12} W/m^2?

SOLUTION:

> **Procedure:** Use the definition of intensity in dB. The first method is to calculate the initial intensity in terms of the reference intensity and then get the new intensity level for a factor of 8. The second method involves using the properties of the logarithm.

First method:
The intensity level, IL, in dB is given in the text by

$$IL = 10 \log_{10}\left(\frac{I}{I_0}\right)$$

where I is the intensity in watts and I_0 is the reference intensity..

We are given that

$$-2.0 \; = \; 10 \times \log_{10}\!\left(\frac{I}{I_0}\right)$$

(1)

and asked to find

$$IL \; = \; 10 \times \log_{10}\!\left(\frac{8I}{I_0}\right).$$

(2)

We can solve equation (1) for I as

$$I \; = \; I_0 \times 10^{-0.2}$$

and substitute this into (2) to get

$$IL \;\; = \;\; 10 \times \log_{10}\!\left(\frac{8 \times 10^{-0.2}I_0}{I_0}\right) \; = \; 10 \times \log_{10}(8 \times 10^{-0.2})$$

$$= \; 10 \times \log_{10}8 + 10 \times \log_{10}10^{-0.2} \; = \; 9.031 + (-2.00) \; = \; 7.0$$

The new power is 7.0 dB.

Second Method:
Use the *log* property that

$$\log\,(a \times b) \; = \; \log a + \log b\,.$$

Subtract (2) from (1) to get

$$IL - (-2.0) \;\; = \;\; 10 \times \log_{10}\!\left(\frac{8.0\,I}{I_0}\right) - 10 \times \log_{10}\!\left(\frac{I}{I_0}\right)$$

$$= \; 10 \times \log_{10}8.0 + 10 \times \log_{10}\!\left(\frac{I}{I_0}\right) - 10 \times \log_{10}\!\left(\frac{I}{I_0}\right)$$

$$= \; 10 \times \log_{10}8 \; = \; 9.0$$

Again $IL \; = \; -2.0 + 9.0 \; = \; 7.0\,\text{dB}.$

Since I_0 is given as $10^{-12}\,\text{W/m}^2$, we can use the first equation to find that

$$I \; = \; I_0 \times 10^{-0.2} \; = \; 6.3 \times 10^{-13}\,\text{W/m}^2.$$

Whenever the intensity is less than the reference level, the relative intensity in dB will be negative.

PROBLEM SG 14.5: A train is approaching an observer at a speed of 60 m/s. A horn of 5,000 Hz on the train sounds as if it has a frequency which is offset 1000 Hz from the original 5,000 Hz. What is the speed of sound at the time of this observation?

SOLUTION:

Procedure: Use the Doppler shift formula for an approaching source.

If we let f be the emitted frequency, f' to observed frequency, v the approach velocity and v_0 the velocity of sound then

$$f' = \frac{f}{1 - \frac{v}{v_0}} \, ,$$

where the observed frequency must be larger than the emitted frequency.

Let Δf be the difference between the emitted and the observed frequency, then

$$f' = = f + \Delta f = \frac{f}{1 - \frac{v}{v_0}} \, .$$

So

$$v = v_0 \times \left(1 - \frac{f}{f + \Delta f}\right) = \frac{v_0 \, \Delta f}{f + \Delta f}$$

and we get

$$v_0 = v \times \frac{f + \Delta f}{\Delta f} \, .$$

Entering the given numbers we get a velocity of sound of 360 m/s.

Chapter 15
Electric Charge and Electric Field

Important Terms

Fill in the blanks with the appropriate word or words.

1. The negative mobile charges often found in solids are called _____ .

2. The principle that allows us to treat the total electrical force on an object as the sum of the individual forces is called the law of _____ .

3. The nuclei of atoms have a positive _____ charge.

4. The law giving the force exerted by one point charge on another is called _____ Law.

5. The _____ is the electric force per unit charge felt by a small test charge.

6. Two equal and opposite charges separated by a small distance are called a(n) _____ .

7. If we write Coulomb's Law in the form

$$\frac{1}{4\pi\varepsilon_0} \frac{q_1 q_2}{r^2},$$

the quantity ε_0 is called the _____ constant.

8. A(n) _____ is a material through which charge can easily flow.

9. The _____ through a small element of surface is the product of the magnitude of the electric field and the projection of the surface perpendicular to the field.

10. Electrons have a negative electric _____.

11. The strength of an electric dipole can be given by either the magnitude of one of the dipole charges involved and the separation of the charges or by the product of these which is called the dipole _____ .

12. A(n) _____ does not permit charges to move freely.

13. The technique of holding a metal object with an insulating handle, bringing it near a charged object and then touching the metal with a finger, removing the finger and then removing the globe is the process of charging _____ .

14. _____ Law is that the net electric flux through any closed surface is directly proportional to the net electric charge enclosed within that surface.

15. Single charges can neither be created nor destroyed is a statement of the _____ .

16. The SI unit of electric charge is the _____.

17. A(n) _____ surface is an imaginary surface constructed to allow us to apply Gauss's Law.

18. The study of electrical charges at rest is called _____ ___.

Write the definitions of the following:

19. dipole moment (15.8):

20. permittivity constant (and value) (15.2):

21. charging by induction (15.1):

22. conductor (15.1):

23. insulator (15.1):

24. electron (15.2):

25. coulomb (15.4):

26. electric charge (15.2):

27. Gaussian surface (15.6):

28. Gauss's Law (15.6):

29. conservation of charge (15.1):

30. Coulomb's Law (15.2):

31. electrostatics (15.0):

32. electric field (15.4):

33. electric flux (15.6):

34. electric dipole (15.8):

Answers to 1–18, Important Terms

1. electrons; 2. superposition; 3. electric; 4. Coulomb's; 5. electric field; 6. electric dipole; 7. permittivity; 8. conductor; 9. electric flux; 10. charge; 11. moment; 12. insulator; 13. by induction; 14. Gauss's; 15. conservation of charge; 16. coulomb; 17. Gaussian; 18. electrostatics

A Word on Models

We have no pure charges. That is, all charged bodies have mass although not all bodies with mass have charge. So we only need add a new property, the charge, to our idealized particles. We idealize an electron to have a mass (m_e = 9.11×10^{-31} kg) and a charge (e = -1.60×10^{-19} C) among other properties, but not to have extension. We will later see that in everyday experience all charges come in multiples of the electron charge. (The exception is quarks in nuclei which may have $-e/3$ or $+2e/3$ as their charge.) The charge on an electron is very small so we can assume that we have test particles with as small a charge as desired. These test particles can be used to investigate the electric fields produced by other charges. In general a test charge is like the test particle used for the gravitational field; that is, a test charge is assumed to have a small enough charge that it does not disturb the positions of the charges we are investigating. Likewise a point particle is one whose size is very small compared with other lengths in the problem under investigation.

All of our models in this chapter do not explain why the negative charges are usually the mobile ones. For now we still assume a solid has no structure other than its shape, mass and electrical conductivity. Solids will be taken as either perfect metals or perfect insulators. Real solids have more complicated properties.

Sign Convention

Coulomb's Law has the sign convention that an attractive force is negative and a repulsive force is positive. This tells you the direction of the force with respect to the line joining the two charges. This does not tell you the direction of the force with respect to an external coordinate system. You will need to draw a picture of the physical situation to determine whether the forces point in a positive or a negative vector sense. Sample Solution SG 15.2 gives you a good example of this problem and how it should be handled.

Sample Solutions

PROBLEM SG 15.1: Two charges exert an attractive force of 6.2×10^{-4} N on each other. The larger charge is twice the magnitude of the smaller. The smaller charge is 2.0×10^{-6} C. What is the larger charge and the separation of the two charges?

SOLUTION:

Procedure: Use Coulomb's law to calculate the forces and solve for the unknowns.

Since the forces are attractive the charges must have opposite sign. If we let q represent the 2.0×10^{-6} C charge, then the other charge must be $-2q$ or -4.0×10^{-6} C. We let d be the separation and F be the attractive force of 6.2×10^{-4} N as shown in Figure SG 15.1.

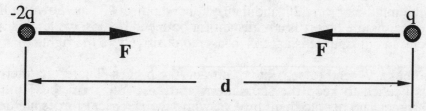

Figure SG 15.1: Charges q and $-2q$ separated by a distance d.

We can now write Coulomb's Law as

$$F = \frac{1}{4\pi\varepsilon_0} \frac{(-2q)(q)}{d^2} = \frac{1}{4\pi\varepsilon_0} \frac{2q^2}{d^2}.$$

We know all the values except the value of d.

GIVEN:

Quantity	Symbol	Value
charge of smaller	q	2.0×10^{-6} C
attractive force	F	6.2×10^{-4} N
separation	d	?
constant	$\dfrac{1}{4\pi\varepsilon_0}$	9.0×10^9 N m^2/C^2

so

$$F = \frac{1}{4\pi\varepsilon_0} \frac{2q^2}{d^2} = \frac{9.0 \times 10^9 \text{ N m}^2/\text{C}^2 \times 2 \times (2.0 \times 10^{-6} \text{ C})^2}{d^2}$$

$$= 6.2 \times 10^{-4} \text{ N}.$$

Thus

$$d^2 = \frac{9.0 \times 10^9 \text{ N m}^2/\text{C}^2 \times 2 \times (2.0 \times 10^{-6} \text{ C})^2}{6.2 \times 10^{-4} \text{ N}} = 11.6 \text{ m}^2.$$

So the other charge is -4.0×10^{-6} C and the two charges are separated by 3.4 meters.

PROBLEM SG 15.2: Two charges of charge $+3.5 \times 10^{-8}$ C and mass 0.10 g are held in position at a separation of 20 cm. Midway between them a charge of -2.0×10^{-8} C and mass 0.20 g is also held. What are the forces needed on each charge to maintain the position of the charges? Assume the experiment was done in a space station so that no gravitational forces are involved. If one of the two outside charges is released, what is its initial acceleration?

SOLUTION:

> **Procedure:** An equilibrium problem — assume the sum of the mechanical plus electrical forces on each particle is zero.

We first need to find the electrostatic forces on each charge. Let us call the left hand object "1", the middle object "2", and the right hand object "3" as shown in Figure SG 15.2a. Since the middle object is equidistant for both end bodies, we let this separation be $d = 10$ cm. In our figure distance is measured positively to the right.

Electrical forces will tend to separate or attract the bodies. Therefore, there must be other forces needed to keep the separations constant. We term these other forces mechanical forces and denote them by F_M, while the electrical forces are denoted by F_E.

We will have to be careful with sign convention when we apply Coulomb's Law. A positive sign means a repulsive force and a negative sign is an attractive force. Consider charges "1" and "2" only. Since these charges are of the opposite sign, the force between them is attractive. This means that the force on charge "1" due to charge "2" is directed to the right and is hence along the $+x$ direction and therefore a positive force in the vector sense. On the other hand, the force on charge "2" due to charge "1" is directed in the $-x$ direction and is therefore a negative force in the vector sense.

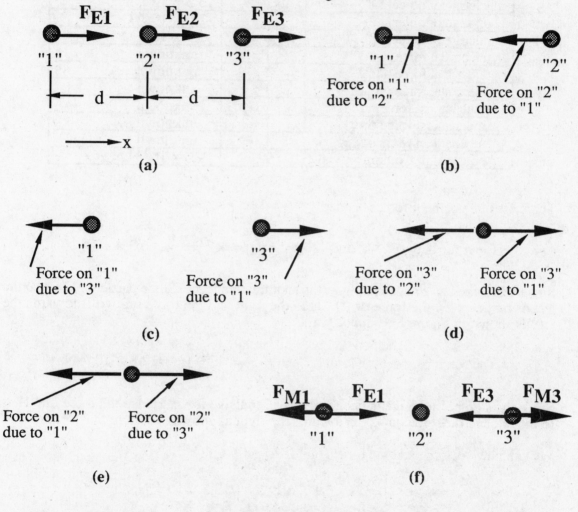

Figure SG 15.2:   a). The three bodies and the total electric force on each shown. The electric forces were arbitrarily shown to the right.
b). The electric forces on "1" and "2" due to each other.
c). The electric forces on "1" and "3" due to each other.
d). The total electric forces on "1".
e) The total electric forces on "2".
f). The total forces needed to keep the system in equilibrium. Electric forces are now shown in their proper orientation.

GIVEN:

Quantity	Symbol	Value
mass of object "1"	m_1	0.10 g
mass of object "3"	$m_3 = m_1$	0.10 g $= 1.0 \times 10^{-4}$ kg
mass of object "2"	m_2	0.20 g $= 2.0 \times 10^{-4}$ kg
separation 1-2, 2-3	d	10 cm $= 1.0 \times 10^{-1}$ m
charge of object "1"	q_1	3.5×10^{-8} C
charge of object "2"	q_2	-2.0×10^{-8} C
charge of object "3"	$q_3 = q_1$	3.5×10^{-8} C
electrical force on "1"	F_{E1}	unknown
electrical force on "2"	F_{E2}	unknown
electrical force on "3"	F_{E3}	unknown
mechanical force on "1"	F_{M1}	unknown
mechanical force on "2"	F_{M2}	unknown
mechanical force on "3"	F_{M3}	unknown
constant in Coulomb's Law	k	9.0×10^9 N m^2/C^2

There are three pairs of forces needed:

$$\text{The force between "1" and "2"} = F_{1\text{-}2} = k\frac{q_1 q_2}{d^2} = -6.3 \times 10^{-4} \text{ N.}$$

Since this force has a negative sign, it is an attractive force. Since particle "1" is the the left of particle "2", the force on "1" is to the right and the force on two is towards the left as shown in Figure SG 15.2b.

$$\text{The force between "1" and "3"} = F_{1\text{-}3} = k\frac{q_1 q_3}{(2d)^2} = 2.8 \times 10^{-4} \text{ N.}$$

Since the sign of this force is positive, it is a repulsive force. Hence the force on "1" is to the left and the force on "2" is towards the right as shown in Figure SG 15.2c.

We can now vectorial add the forces on "1" to get

$$F_{E1} = 6.3 \times 10^{-4} \text{ N} - 2.8 \times 10^{-4} \text{ N} = 3.5 \times 10^{-4} \text{ N.}$$

This force is positive since it is directed towards the right (in a positive x-direction)

A repeat of our analysis for "3" must, by symmetry, give us the same numerical value but the opposite direction, so, as shown in Figure SG 15.2d,

$$F_{E3} = -6.3 \times 10^{-4} \text{ N} + 2.8 \times 10^{-4} \text{ N} = -3.5 \times 10^{-4} \text{ N.}$$

Finally F_{2-1} has the same magnitude as F_{2-3} but is in the opposite direction as shown in Figure SG 15.2e

$$F_{E2} = -6.3 \times 10^{-4} \text{ N} + 6.3 \times 10^{-4} \text{ N} = 0.$$

These three forces are shown in Figure SG 15.2f, pointing in the proper directions.

The mechanical forces needed to maintain the configuration are also shown in Figure SG 15.2f. They have the same magnitude but are in the opposite direction to the electric forces.

$$F_{M1} = -F_{E1}; \qquad F_{M2} = -F_{E2} = 0; \qquad F_{M3} = -F_{E3}.$$

If the left outside charge were released, the initial acceleration would be given by

$$\text{acceleration} = \frac{\text{Force}}{\text{mass}} = \frac{F_{E1}}{m_1} = 3.5 \text{ m/s}^2.$$

The acceleration is positive, indicating that particle "1" is accelerated in the positive x-direction.

PROBLEM SG 15.3: A negative charge of 2.7×10^{-7} C is held in a uniform electric field of 1.0×10^3 V/m. Another negative charge of 7.2×10^{-7} C is brought to a distance d from the first charge. What is the value of d such that that the force of either charge on the other is equal in magnitude to the force on the first charge due to the uniform field?

SOLUTION:

Procedure: Use Coulomb's Law to calculate the electric field due to the second negative charge.

For the purposes of a drawing, we imagine the electric field is directed to the left as shown in Figure SG 15.3. We will let q_1 be the first charge and q_2 be the second. To stress the nature of the question, the two particles have not been placed along the direction of the field.

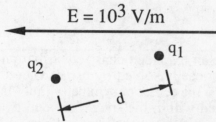

Figure SG   15.3:   Two charges a distance d apart in a uniform electric field E.

GIVEN:

Quantity	Symbol	Value
uniform electric field	E	1.0×10^3 V/m
first charge	q_1	2.7×10^{-7} C
second charge	q_2	-7.2×10^{-7} C
separation of charges	d	unknown
constant in Coulomb's Law	k	9.0×10^9 N m^2/C^2

We are asked to choose d so that the magnitude of the force on "1" due to the field is the same as the magnitude of the force on "1" due to "2".

The force on "1" due to the field is

$$F_E = q_1 E = -2.7 \times 10^{-7}\, C \times 1.0 \times 10^3\, V/m = -2.7 \times 10^{-4}\, N$$

where the minus sign indicates the force is in a direction opposite to E (i.e. to the right).

The force on "1" due to "2" is

$$F_C = k \frac{q_1 q_2}{d^2} = \frac{1.7 \times 10^{-3}\, N\text{-}m^2}{d^2}$$

where the positive sign indicates that the force is repulsive.

Equating the magnitudes of these two forces gives

$$2.7 \times 10^{-4}\, N = \frac{1.7 \times 10^{-3}\, N\text{-}m^2}{d^2},$$

or

$$d = \sqrt{6.3\, m^2} = 2.5\, m.$$

The second charge must be placed 2.5 m from the first charge.

PROBLEM SG 15.4: Gauss's Law I. A spherical shell has an inner radius of 0.25 m and an outer radius of 0.30 m. If a total charge of 2.0×10^{-3} C is placed on the shell, give the electric field as a function of distance from the center of the shell if the shell is a perfect conductor.

SOLUTION:

> **Procedure:** Since we have an extended charge distribution with high symmetry, we expect this problem is most easily solved by use of Gauss's Law. The electric field crossing a sphere centered on the center of symmetry times the area of that sphere must be the total charge contained within the sphere divided by ε_0. Choose the spherical surface to be inside the shell, within the shell and outside the shell.

Since this is a problem with spherically symmetry, we expect the electric field will depend only upon the the distance from the center of the sphere; and that any electric field present must be radially directed. Since we are not dealing with point charges, the only tool that seems reasonable to use is Gauss's Law (text section 15.7). This law states that the flux through a surface is proportional to the net charge, Q, contained within the surface, i.e.

$$\Phi_E = \frac{Q}{\varepsilon_0}.$$

The simplest choice for the Gaussian surface to use is to take a sphere of radius r, centered on the same center used for the spherical shell. Other choices can be made, but the mathematics are above the level of this course, and the results are the same.

With this choice we have

$$\Phi_E(r) = E(r) \times (\text{area of sphere of radius } r)$$

$$= E(r) \times 4\pi \, r^2.$$

Gauss's Law becomes

$$E(r) = \frac{Q}{4\pi \, \varepsilon_0 \, r^2} \, ,$$

where Q is the net charge contained within r, and

$$\frac{1}{4\pi \, \varepsilon_0} = 9.0 \times 10^9 \text{ N-m}^2/\text{C}^2.$$

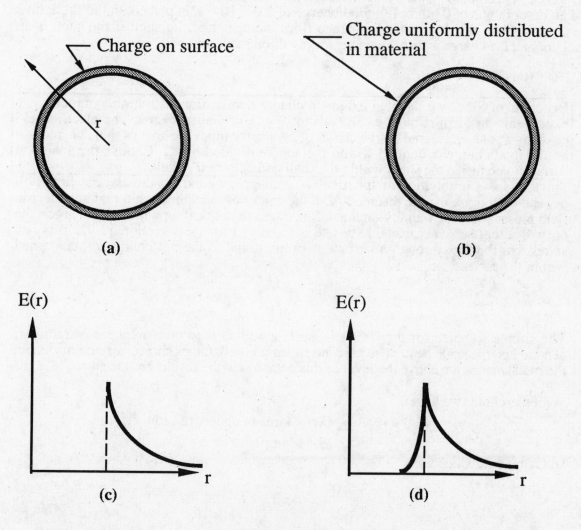

Figure SG 15.4:   a). Charge on the surface of a conducting spherical shell.
b). Charge distributed uniformly in a spherical shell.
c). The electric field for case a.
d). The electric field for case b.

For a perfect conductor, the entire charge must lie on the outer surface of the shell (see section 15.6), so

$$E(r) = \begin{cases} 0 & r < 0.30 \text{ m} \\ \dfrac{1.8 \times 10^7 \text{ N-m}^2/\text{C}}{r^2} & r \geq 0.30 \text{ m} \end{cases}.$$

That is, the electric field is zero within the outer surface of the shell and falls off as the square of the distance from the center for distances outside the shell. This behavior is shown in Figure SG 15.4c.

PROBLEM SG 15.5: Gauss's Law II. A spherical shell has an inner radius of 0.25 m and an outer radius of 0.30 m. If a total charge of 2.0×10^{-3} C is placed on the shell, give the electric field as a function of distance from the center of the shell if the shell is an insulator and the charge is distributed uniformly throughout the shell.

SOLUTION:

> **Procedure:** Since we have an extended charge distribution with high symmetry, we expect this problem is most easily solved by use of Gauss's Law. The electric field crossing a sphere centered on the center of symmetry times the area of that sphere must be the total charge contained within the sphere divided by ε_0. Choose the spherical surface to be inside the shell, within the shell and outside the shell.

Since this is a problem with spherically symmetry, we expect the electric field will depend only upon the the distance from the center of the sphere; and that any electric field present must be radially directed. Since we are not dealing with point charges, the only tool that seems reasonable to use is Gauss's Law (text section 15.7). This law states that the flux through a surface is proportional to the net charge, Q, contained within the surface, i.e.

$$\Phi_E = \frac{Q}{\varepsilon_0}.$$

The simplest choice for the Gaussian surface to use is to take a sphere of radius r, centered on the same center used for the spherical shell. Other choices can be made, but the mathematics are above the level of this course, and the results are the same.

With this choice we have

$$\Phi_E(r) = E(r) \times (\text{area of sphere of radius } r)$$
$$= E(r) \times 4\pi \, r^2,$$

so Gauss's Law becomes

$$E(r) = \frac{Q}{4\pi \, \varepsilon_0 \, r^2},$$

where Q is the net charge contained within r, and

$$\frac{1}{4\pi \, \varepsilon_0} = 9.0 \times 10^9 \text{ N-m}^2/\text{C}^2.$$

For this problem the total charge contained inside r is

and

$$\text{zero for } r < 0.25 \text{ m}$$

$$2.0 \times 10^{-3} \text{ C for } r \geq 0.30 \text{ m;}$$

and, the charge for $0.25 \text{ m} \leq r \leq 0.30 \text{ m}$ will vary with r.

Now for $r < 0.25 \text{ m}$

$$E(r) = \frac{Q}{4\pi \, \varepsilon_0 \, r^2} = \frac{0.0 \text{ C}}{4\pi \, \varepsilon_0 \, r^2} = 0,$$

while for $r > 0.30 \text{ m}$

$$E(r) = \frac{Q}{4\pi \, \varepsilon_0 \, r^2} = \frac{2.0 \times 10^{-3} \text{ C}}{4\pi \, \varepsilon_0 \, r^2} = \frac{1.8 \times 10^7 \text{ N-m}^2/\text{C}}{r^2}$$

We now examine the behavior for $0.25 \text{ m} < r < 0.30 \text{ m}$. What we need is the total amount of charge contained within r. Since the charge is distributed uniformly within the shell, the total amount of charge within r will be proportional to the volume of the shell within r. The total volume between $0.25 \text{ m} = r_{in}$ and $0.30 \text{ m} = r_{out}$ can be calculated by subtracting the volume of a sphere of radius r_{in} from the volume of a sphere of radius r_{out}, giving:

$$\text{Total Volume} = \frac{4\pi}{3} \, r_{out}^3 - \frac{4\pi}{3} \, r_{in}^3 = \frac{4\pi}{3} \, (r_{out}^3 - r_{in}^3),$$

while the volume between $0.25 \text{ m} = r_{in}$ and some $r \leq r_{out}$, is

$$\text{Volume at r} = \frac{4\pi}{3} r^3 - \frac{4\pi}{3} r_{in}^3 = \frac{4\pi}{3} (r^3 - r_{in}^3).$$

We can find the charge inside r by a proportion

$$\frac{\text{charge within } r}{\text{total charge}} = \frac{\text{volume (inside } r \text{) containing charge}}{\text{total volume containing charge}}$$

or

$$\frac{Q(r)}{Q} = \frac{Q(r)}{2.0 \times 10^{-3} \text{ C}} = \frac{\text{Volume at } r}{\text{Total Volume}}$$

$$= \frac{\frac{4\pi}{3} (r^3 - r_{in}^3)}{\frac{4\pi}{3} (r_{out}^3 - r_{in}^3)} = \frac{r^3 - r_{in}^3}{r_{out}^3 - r_{in}^3}.$$

This gives for $0.25 \text{ m} \leq r \leq 0.30 \text{ m}$

$$Q(r) = 2.0 \times 10^{-3} \text{ C} \, \frac{r^3 - (0.25 \text{ m})^3}{(0.30 \text{ m})^3 - (0.25 \text{ m})^3}$$

$$= 0.18 \frac{\text{C}}{\text{m}^3} \times [(0.30 \text{ m})^3 - (0.25 \text{ m})^3].$$

$$E(r) = \kappa \frac{Q(r)}{r^2},$$

hence for all r

$$E(r) = \begin{cases} 0 & r < 0.25\,\text{m} \\ 1.6 \times 10^9 \,\frac{\text{N}}{\text{C-m}} \left(r - \frac{(0.25\,\text{m})^3}{r^2}\right) & 0.25\,\text{m} \le r \le 0.30\,\text{m} \\ \dfrac{1.8 \times 10^7 \,\text{N-m}^2/\text{C}}{r^2} & r > 0.30\,\text{m} \end{cases}$$

This field is shown in Figure SG 15.4d. At the boundaries you must change the formulae used to calculate $E(r)$. You can easily verify that each formula gives the same value at the boundaries. Note that the electric field, if it is not trivially zero, must depend
upon the distance r from the center of symmetry.

PROBLEM SG 15.6: Gauss's Law III. A point charge of 2.0×10^{-8} C is located at the center of an uncharged conducting spherical shell of inner radius 0.15 m and outer radius 0.20 m. What is the electric field inside the shell, within the surfaces of the shell and outside the shell? What charges are induced on the surfaces of the shell?

SOLUTION:

Procedure: Since we have an extended charge distribution with high symmetry, we expect this problem is most easily solved by use of Gauss's Law. The electric field crossing a sphere centered on the center of symmetry times the area of that sphere must be the total charge contained within the sphere divided by ε_0. Choose the spherical surface to be inside the shell, within the shell and outside the shell.

Reread the first parts of Problems SG 15.4 and 15.5. For spherical symmetry the electric field, $E(r)$ is given by

$$E(r) = \frac{Q}{4\pi\,\varepsilon_0\,r^2},$$

where Q is the net charge contained within r, and

$$\frac{1}{4\pi\,\varepsilon_0} = 9.0 \times 10^9 \,\text{N-m}^2/\text{C}^2.$$

There are once again three regions to analyze: $0 < r < 0.15$ m, 0.15 m $< r < 0.20$ m, and $r > 0.20$ m. We also know that the presence of the point charge may cause charge to be induced on the inner and outer surface of the conducting shell. We will denote these charges by Q_{in} and by Q_{out} respectively; and note that $Q_{in} + Q_{out} = 0$ since there in no net charge on the sphere. We can now make a table of $Q(r)$ for the various regions:

radius	Total charge contained
$0 < r < 0.15$ m	2.0×10^{-8} C
0.15 m $< r < 0.20$ m	2.0×10^{-8} C $+ Q_{in}$
0.20 m $< r$	2.0×10^{-8} C $+ Q_{in} + Q_{out}$

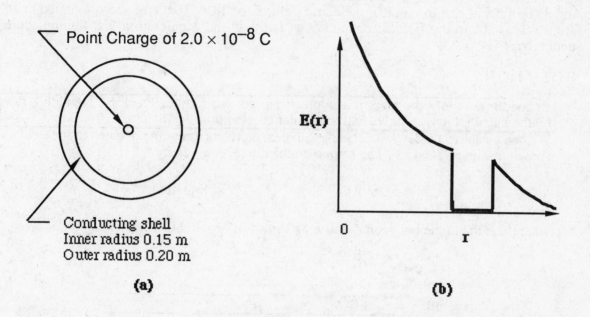

Figure SG 15.5: a) A point charge is located inside a hollow spherical conducting sphere. b). The resulting electric field.

Since $Q_{in} + Q_{out} = 0$, we know $Q(r)$ for $r > 0.20$ m, and hence the electric fields for $r < 0.15$ m and for $r > 0.20$ m are known and are the same form in each region since Q is constant in each region. Thus

$$E(r) = \frac{Q}{4\pi\,\varepsilon_0\,r^2} = \frac{1.8 \times 10^2 \text{ N-m}^2}{r^2} \quad \begin{cases} r < 0.15 \text{ m} \\ r > 0.20 \text{ m} \end{cases}$$

It remains then to determine Q_{in} and the electric field for 0.15 m $< r <$ 0.20 m. But this range of r is inside the conductor. We know that no electric field exists inside a conductor so $E(r)$ for 0.15 m $< r <$ 0.20 m must be zero; and, hence the induced charge $Q_{in} = -2.0 \times 10^{-8}$ C on the inner surface and $Q_{out} = 2.0 \times 10^{-8}$ C on the outer surface.

$$E(r) = 0 \quad 0.15 \text{ m} < r < 0.20 \text{ m}.$$

Note that in this problem the electric field is not continuous across the surfaces. This should be expected when ever there is a surface charge distribution present.

PROBLEM SG 15.7: A molecule of HCl, which has an electric dipole moment p = 3.4×10^{-33} C-m and moment of inertia about its center of mass of $I = 2.65 \times 10^{-47}$ kg-m^2, is initially at rest in a uniform electric field, E, of 1.0×10^3 V/m and orientated at an angle, θ, of 30° to the field direction. What is the initial acceleration that the molecule undergoes?

SOLUTION:

> **Procedure:** Apply the electric torque law to get the torque. Then use Newton's law relating applied torque and resulting angular acceleration.

We can visualize the HCl molecule as an ideal dipole of strength p located in the field as shown in Figure SG 15.6. The torque on the dipole is given by

$$\tau = pE \sin(\theta), \qquad \text{(text Equation 15.4)}$$

where θ is the angle between P and E as shown in Figure SG 15.6.

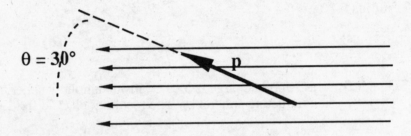

$\theta = 30°$ $\qquad$ p

Figure SG 15.6: Water molecule in electric field.

The law relating angular acceleration and torque was given in Chapter 7 of the text as

$$\tau = I \alpha, \qquad \text{(text Equation 7.17)}$$

where I is the moment of inertia and α is the angular acceleration. Thus

$$\alpha = \frac{pE \sin(\theta)}{I}.$$

GIVEN:

Quantity	Symbol	Value
dipole moment	p	3.4×10^{-33} C m
moment of inertia	I	2.65×10^{-47} kg m^2
electric field strength	E	1.0×10^3 V/m
angle between p and E	θ	30°
angular acceleration	α	unknown

So the angular acceleration will be 6.4×10^{16} rad/s^2 in a direction tending to reduce θ (counterclockwise in the figure).

Chapter 16
Electric Potential and Capacitance

Important Terms

Fill in the blanks with the appropriate word or words.

1. The _____ of a material is the ratio of the capacitance of a capacitor with this material between the plates to the capacitance of the same capacitor with a vacuum between the plates.

2. The ratio of the change in electric potential to the distance we move is called the _____ in that direction.

3. The maximum potential per unit thickness that a material can withstand without breakdown is called the _____ of the material.

4. The _____ between two points is defined as the ratio of the electric potential energy change that occurs when a test charge is moved between the two points to the value of the charge.

5. The _____ of a capacitor is the ratio of the charge stored to the potential difference.

6. _____ accelerators are machines used to accelerate charged particles to very high kinetic energies.

7. The _____ is the SI unit of electric potential.

8. Materials which do not permit the conduction of electric charge are called insulators or _____ .

9. A surface of constant potential is called a(n) _____ .

10. A(n) _____ is a device to store charge.

11. The unit of capacitance is the _____, it is 1 Coulomb per Volt.

12. _____ energy is the ability to do work stored in the electric field.

Write the definitions of the following:

13. electric potential (16.1):

14. equipotential surface (16.3):

15. volt (16.1):

16. dielectric strength (16.8):

17. Van de Graaff accelerator (16.2):

18. dielectric (16.8):

19. potential gradient (16.4):

20. capacitor (16.5):

21. capacitance (16.5):

22. farad (16.5):

23. electric potential difference (16.1):

24. dielectric constant (16.8):

Answers to 1–12, Important Terms

1. dielectric constant; 2. potential gradient; 3. dielectric strength; 4. electric potential; 5. capacitance; 6. Van de Graaff; 7. coulomb; 8. dielectrics; 9. equipotential surface; 10. capacitor; 11. farad; 12. electric potential

Motion in the Electric Field

The electric field is defined to go from high potential to low potential. This direction is defined by the way a positive charge will move if it is released in the field. This charge is a test charge — that is, a test charge has a charge is sufficiently small that we can neglect the effects of the test charge on the electric fields present. Remember that positively charged particles feel a force in the same direction that the field lines point while negatively charged particles feel a force in the opposite direction.

Sample Solutions

PROBLEM SG 16.1: Two charges of charge $+3.5 \times 10^{-8}$ C and mass 0.10 g are held in position at a separation of 20 cm. Midway between them a charge of -2.0×10^{-8} C and mass 0.20 g is also held. What is the electric field seen by each charge What is the electric potential at a point midway between the first and second charge?

SOLUTION:

Procedure: Use the definition of electric field and vectorially sum the individual fields. (See Figure SG 15.2 for the definitions of the terms used.)

This is Problem SG 15.2 of the last chapter, so we can use the the figure for and the results of that problem. In terms of Figure SG 15.2, the force on the left hand charge is

$$F_{E1} = 3.5 \times 10^{-4} \text{ N},$$

the force on the right hand charge is

$$F_{E3} = -3.5 \times 10^{-4} \text{ N},$$

and the central charge feels a force

$$F_{E2} = 0 \text{ N}.$$

The sign convention is positive towards the right.

The electric field is defined as

$$E = \frac{F}{q},$$

where the sign of the charge, q, must be included. That is, if the charge is negative the electric field will point in the opposite direction to the force.

So the electric fields are

$$E_{E1} = \frac{F_{E1}}{q_1} = 1.0 \times 10^5 \text{ N/C} = 1.0 \times 10^5 \text{ V/m},$$

$$E_{E2} = 0,$$

and

$$E_{E3} = -1.0 \times 10^5 \text{ V/m}.$$

The potential at any point along the line joining the charges is given by

$$V = \frac{q_1}{4\pi\varepsilon_0 \, r_1} + \frac{q_2}{4\pi\varepsilon_0 \, r_2} + \frac{q_3}{4\pi\varepsilon_0 \, r_3},$$

where r_i is the distance from the point in question to the ith charge as is shown in Figure SG 16.1. That is, r_1 is the distance between the choosen point and the charge named "1", r_2 is the distance between the choosen point and the charge marked "2", and r_3 is the distance between the choosen point and the charge named "3".

This expression for force is correct only when all the charges are in a line and then only for points along that line.

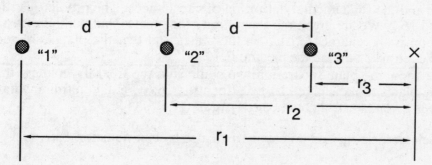

Figure SG 16.1: The field point located by an "×" relative to the three charges. This figure is drawn for a field point to the right of all three charges. The question has the field point midway between charges "1" and "2".

GIVEN:

Quantity	Symbol	Value
charge of object "1"	q_1	3.5×10^{-8} C
charge of object "2"	q_2	-2.0×10^{-8} C
charge of object "3"	$q_3 = q_1$	3.5×10^{-8} C
distance from q_1	r_1	5.0 cm = 0.050 m
distance from q_2	r_2	5.0 cm = 0.050 m
distance from q_3	r_3	15 cm = 0.15 m
constant in Coulomb's Law	$1/(4\pi\varepsilon_0)$	9.0×10^9 N m^2/C^2

So, at the point in question,

$$V = \frac{3.5 \times 10^{-8} \text{ C} \times 9.0 \times 10^9 \text{ N m}^2/\text{C}^2}{0.050 \text{ m}}$$

$$+ \frac{-2.0 \times 10^{-8} \text{ C} \times 9.0 \times 10^9 \text{ N m}^2/\text{C}^2}{0.050 \text{ m}}$$

$$+ \frac{3.5 \times 10^{-8} \text{ C} \times 9.0 \times 10^9 \text{ N m}^2/\text{C}^2}{0.150 \text{ m}} ,$$

giving,

$$V = 4.8 \times 10^3 \text{ N m/C} = 4.8 \times 10^3 \text{ V}.$$

So midway between the two charges the potential is about 4.8×10^3 Volts with respect to the potential at a large distance from the charges.

Problem SG 16.2: A pair of parallel plates is separated by 1.672 mm of Mylar. There is a charge of 2.52×10^{-8} C/m^2 on each plate (positive on the upper in Figure SG 16.1 and negative on the lower. What is the electric field between the plates neglecting edge effects?

SOLUTION:

Procedure: Use the definition of capacitance for a parallel plate capacitor with a dielectric and the fundamental definition of capacitance. The problem is illustrated in Figure SG 16.2. We are given the separation of the plates, but not their area. The other information is the charge density on the plates and that the plates are separated by Mylar. We are asked to find the electric field.

Since we know the plate separation, we could find the field if we knew the potential across the plates. This is given by text Equation 16.6. First, determine what is needed and what is given (see table on the next page).

The capacitance, charge and potential are related by text Equation 16.7

$$V = \frac{q}{C} = \frac{\rho A}{\kappa\varepsilon_0 \dfrac{A}{d}} = \frac{\rho d}{\kappa\varepsilon_0} ,$$

so the electric field is given by

$$E = -\frac{V}{d} = -\frac{\rho}{\kappa\varepsilon_0}$$

$$= -\frac{2.53 \times 10^{-8}\ \text{C/m}^2}{3.1 \times 8.85 \times 10^{-12}\ \text{C}^2/\text{N-m}^2} = -9.2 \times 10^2\ \text{V/m.}$$

1.672 mm

Mylar

Electric field $= 2.52 \times 10^{-8}$ C/m^2

Figure SG 16.2: A parallel plate capacitor with a Mylar dielectric.

GIVEN:

Quantity	Symbol	Value
Area of plates	A	unknown
separation of plates	d	1.672 mm
dielectric constant (Table 16.2)	κ	3.1
relation for capacitance	C	$\kappa\varepsilon_0\dfrac{A}{d}$
charge density	ρ	2.53×10^{-8} C/m^2
total charge	q	ρA (unknown)
potential across plates	V	unknown
electric field between plates	E	$-\dfrac{V}{d}$ (unknown)

Note that the electric field is independent of the plate separation for a given charge on the plates. The minus sign tells us that the electric field points in the opposite direction to the change in the potential.

PROBLEM SG 16.3: A pair of parallel plates, each of area 250 cm^2, is separated by an air gap of 2.00 mm. An electric field of 5.60×10^4 V/m exists between the plates. How much work must be done to separate the plates so the air gap is 2.50 mm? Neglect the effect of the edges of the plates on the electric field.

SOLUTION:

> **Procedure:** We are asked to find the work ; this suggests consideration of conservation of energy. That idea gives "work done = final energy – initial energy". We need the energy of capacitor which is given by text equation 16.2 as
>
> $$\text{energy} = \frac{1}{2}CV^2$$
>
> So we need both initial and final C (capacitance) and V (voltage) to get the energy. How do we get C? — use text equation 16.12 — can clearly calculate C How to get V? — initial V — given electric field and separation so can find initial V. final V — not obviously given — must calculate — How? What is conserved during displacement? — charge on plates! So use text equation 16.7 to get relation among charge, capacitance and voltage. Use the initial voltage and capacitance to get the charge and then find the final voltage.

The problem is illustrated in Figure SG 16.3.

We will solve this problem using conservation of energy. That is, the initial energy stored in the capacitor plus the work done is the final energy in the capacitor. The energy stored in a capacitor, with capacitance C charged to V volts, is given by text Equation 16.13 as

$$\text{Energy stored} = \frac{1}{2} C V^2$$

while the capacitance of a parallel plate capacitor of area A and plate separation d is given by text Equation 16.8 as

$$C = \varepsilon_0 \frac{A}{d}.$$

Since we know the initial electric field, we can calculate the initial potential as

$$V_{\text{init}} = E_{\text{init}} \times d_{\text{init}}$$

$$= 5.6 \times 10^4 \text{ V/m} \times 2.00 \times 10^{-3} \text{ m} = 112 \text{ V}.$$

The initial capacitance is

$$C_{\text{init}} = \varepsilon_0 \frac{A_{\text{init}}}{d_{\text{init}}}$$

$$= 8.85 \times 10^{-12} \text{ C}^2/\text{N–m} \times \frac{2.50 \times 10^{-2} \text{ m}^2}{2.00 \times 10^{-3}\text{m}}$$

$$= 1.11 \times 10^{-10} \text{ C}^2/\text{N} = 1.11 \times 10^{-10} \text{ F},$$

so the initial electrical energy is

$$\text{Energy}_{\text{init}} = \frac{1}{2} C_{\text{init}} V_{\text{init}}^2$$

$$= 6.96 \times 10^{-7} \text{ N m}^2 = 6.96 \times 10^{-7} \text{ J}.$$

In order to calculate the work done, we will need the final energy of the capacitor which means we need the potential after the plates are separated. To calculate this we will need to find the total charge on the plates using the definition of capacitance in text Equation 16.7, which is

$$q = CV$$

Since no charge is added

$$q_{\text{final}} = q_{\text{init}} = C_{\text{init}} V_{\text{init}} = 1.24 \times 10^{-8} \text{ C}.$$

The final capacitance is

$$C_{\text{final}} = \varepsilon_0 \frac{A_{\text{final}}}{d_{\text{final}}} = C_{\text{init}} \times \frac{d_{\text{init}}}{d_{\text{final}}} = 8.85 \times 10^{-11} \text{ F}.$$

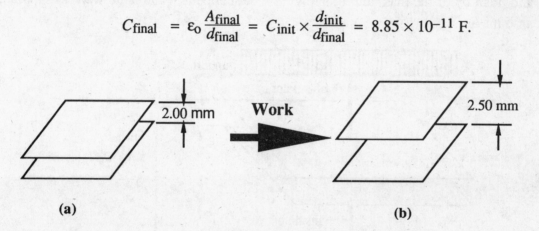

(a)   **(b)**

Figure SG 16.3:   a). The capacitor with initial plate separation.
b). The capacitor with final plate separation.

Thus the final voltage is given by

$$V_{\text{final}} = \frac{q_{\text{final}}}{C_{\text{final}}} = 140 \text{ V},$$

and the final electrical energy

$$\text{Energy}_{\text{final}} = \frac{1}{2} C_{\text{final}} V_{\text{final}}^2 = 8.67 \times 10^{-7} \text{ J}.$$

The work done, W, is

$$W = \text{Energy}_{\text{final}} - \text{Energy}_{\text{init}} = 1.71 \times 10^{-7} \text{ J}$$

It will take 1.71×10^{-7} J of work to separate the plates an additional 0.50 mm.

PROBLEM SG 16.4: A parallel plate capacitor is hung by the center of one plate using a non-conducting wire and another non-conducting wire is attached to the bottom plate. The dielectric between the plates is sticky and is just able to hold the capacitor together, along with the wires, if the capacitor is uncharged. The capacitor is now charged to a potential of 500 V, and the charging source removed. What is the maximum mass that can be hung on the lower wire without pulling the capacitor apart? Hint: Consider the potential energy gain due to the mass of the plate as the plate separation increases and neglect the kinetic energy as the separation first occurs.

SOLUTION:

> **Procedure:** As the plates separate we must do work. Gravitational energy will be gained by the plate moving downward. If the new total energy is less than the original energy, the plate will fall.

The problem is illustrated in Figure SG 16.4 with a) showing the initial plate separation and position of the mass and b) showing what happens if the plate were to separate an amount Δd.

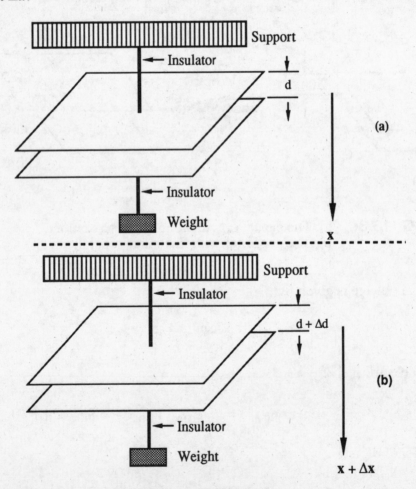

Figure SG 16.4: a). Initial arrangement with plates a distance d apart and the mass added to weight.
b). After the plates have moved an extra distance Δd apart and the mass has moved to $x + \Delta x$.

It takes work to separate the plates of a charged capacitor. Work is done by a mass moving down in a gravitational field. This process will occur if the work provided by moving the mass, m, a distance $\Delta x = \Delta d$ in the gravitational field is greater than the work needed to separate the plates by a distance Δd.

The energy stored in the capacitor is given by text Equation 16.13 as

$$W = \frac{1}{2} CV^2 \,,$$

where C is the capacitance and V the voltage.

Although the capacitor is initially charged to 500 V, this potential will not be constant if the plates separate. What is constant? — The charge on the plates is constant, so text Equation 16.7 should be used to relate the capacitance C and the voltage V. That is,

$$W = \frac{1}{2}CV^2 = \frac{1}{2} C \frac{q^2}{C^2}$$

$$= \frac{q^2}{2C} = \frac{q^2 d}{2\kappa\varepsilon_0 A}$$

where q is the initial charge on the plates, d the plate separation, A the plate area and κ is the dielectric constant. The change in energy, ΔW if the plates separate by Δd, is then

$$\Delta W = \frac{q^2 \Delta d}{2\kappa\varepsilon_0 A}.$$

The energy gained by the mass moving Δx is

$$\Delta E = -mg\Delta x = -mg\Delta d.$$

The total energy change is given by,

$$\Delta U = \Delta W + \Delta E$$

$$= \left(\frac{q^2}{2\kappa\varepsilon_0 A} - m\,g \right).$$

This change is positive provided

$$m < \frac{q^2}{2\kappa\varepsilon_0 A g}.$$

If the mass is larger than the critical value, $\dfrac{q^2}{2\kappa\varepsilon_0 A g}$, the plates will separate.

PROBLEM SG 16.5: A air gap parallel plate capacitor is constructed of two plates of surface area 100 cm² and separation 0.0500 cm. The capacitor is oriented with the plates horizontal in the laboratory which is on earth. The capacitor is charged to a potential of 700.0 V with the upper plate at the positive potential. A physically small charge of mass 0.300 g and charge -1.00×10^{-8} C is placed in the middle of the capacitor and midway between the plates. Describe the motion of the charge. Assume the presence of the charge between the plates does not affect the distribution of charges on the capacitor plates.

SOLUTION:

> **Procedure:** Calculate the electric field inside the capacitor and the resulting electric force on the charge. Vectorially add this force to the gravitational force and determine the resulting motion.

The problem is shown edge on in Figure SG 16.5a. We show the upper plate being at a potential of +700 V with respect to the bottom plate and measure position positively upward from the bottom plate. The charge feels two forces, the electric force and the gravitational force as shown in Figure SG 16.5b.

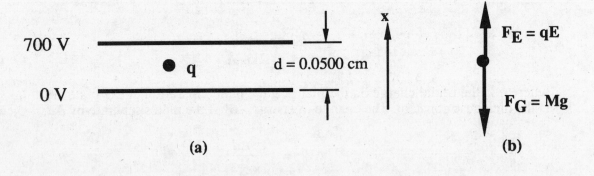

(a) (b)

Figure SG 16.5: a). The physical arrangement of a charge between the charged plates of a capacitor.
b). The forces acting on the charge.

To solve this problem we need to calculate the electric field between the plates and thus find the electric force on the charge. Since we are given the electric potential, we do not need to find the value of the capacitance; that is, the electric field depends only upon the potential difference and the separation of the plates.

GIVEN:

Quantity	Symbol	Value
potential difference	V	700 V
plate separation	d	5.00×10^{-4} m
charge of object	q	-1.00×10^{-8} C
mass of object	M	3.00×10^{-4} kg
acceleration of gravity	g	-9.80 m/s²
electric field	$E = -\Delta V/\Delta x$	unknown
initial position of charge	x_0	2.5×10^{-4} m
position of charge	$x(t)$	unknown

We first find the electric field by the relation

$$E = -\frac{V}{d} = -\frac{700 \text{ V}}{5.00 \times 10^{-4} \text{ m}} = -1.40 \times 10^6 \text{ V/m}.$$

The minus sign tells us that the electric field points in the negative x-direction — i.e. downwards. If we let F_e be the electric force and F_g be the gravitational force, then

$$F_e = qE = -1.00 \times 10^{-8} \text{ C} \times (-1.40 \times 10^6 \text{ V/m}) = 1.40 \times 10^{-2} \text{ N},$$

and

$$F_g = Mg = 3.00 \times 10^{-4} \text{ kg} \times (-9.80 \text{ m/s}^2) = -2.94 \times 10^{-3} \text{ N}.$$

The acceleration, a is given by Newton's Second Law,

$$a = \frac{\text{sum of forces}}{M} = \frac{1.40 \times 10^{-2} \text{ N} + (-2.94 \times 10^{-3} \text{ N})}{3.00 \times 10^{-4} \text{ kg}}$$

$$= 36.9 \text{ N/kg} = 36.9 \text{ m/s}^2.$$

The force and hence the acceleration is positive (that is, upward).

The position is given by the relation

$$x(t) = x_0 + v_0 t + \frac{1}{2} a t^2,$$

where x_0 is the position at $t = 0$ and v_0 is the velocity at $t = 0$. We are measuring the position positive upwards from the bottom plate. The position is then

$$x(t) = 2.5 \times 10^{-4} \text{ m} + 18.5 \text{ m/s}^2 \times t^2.$$

This upward motion can continue until $x(t_f) = 5.00 \times 10^{-4}$ m when the charge hits the plate. This defines $t_f = 3.67 \times 10^{-3}$ s.

The charge will move upward, accelerating at 18.5 m/s^2 until it hits the upper plate in 3.67×10^{-3} seconds. When the charge gets near the upper plate it will cause a redistribution of the charges on the plate and hence the electric field will be slightly different than the field assumed for an ideal parallel plate capacitor so our answer is not exact..

PROBLEM SG 16.6: How much more charge can a given parallel plate capacitor hold without breakdown if the dielectric is Teflon rather than paper?

SOLUTION:

Procedure: The capacitance will change if we change dielectrics. We need to rewrite the relations to eliminate the voltage by using the electric field.

We are asked to compare two parallel plate capacitors which are identical except for the dielectric. The two capacitors are shown in Figure SG 16.6. We let A be the area of the plates and d the plate separation. The breakdown strength of each substance will be denoted by $_{max}E_{material}$. We will also need the dielectric constants. Both are given in text Table 16.2.

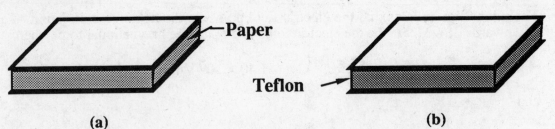

Figure SG 16.6: a) Paper filled capacitor. b) Teflon filled capacitor.

GIVEN:

Quantity	Symbol	Value
dielectric constant of paper	κ_{paper}	3.5
dielectric constant of Teflon	κ_{Teflon}	2.1
breakdown strength of paper	$_{max}E_{paper}$	14×10^6 V/m
breakdown strength of Teflon	$_{max}E_{Teflon}$	60×10^6 V/m

For a capacitor the following equations hold:

$$q = CV \qquad C = \kappa\varepsilon_0 A/d,$$

where C is the capacitance, q the charge on either plate, and V the voltage. We want the maximum charge, given by

$$q_{max} = CV_{max},$$

where the maximum voltage before breakdown can be expressed in terms of the breakdown strength by

$$V_{max} = {_{max}E}\, d.$$

Putting these relations together gives

$$q_{max} = \kappa\varepsilon_0\frac{A}{d} \times {_{max}E}\, d$$

$$= {_{max}E}\, \kappa\varepsilon_0 A.$$

So the maximum charge possible is directly proportional to the dielectric constant and the breakdown strength.

Finally, we get,

$$\frac{\text{maximum charge with Teflon}}{\text{maximum charge with paper}} = \frac{\text{max}E_{\text{Teflon}}\ \kappa_{\text{Teflon}}}{\text{max}E_{\text{paper}}\ \kappa_{\text{paper}}} = 2.6.$$

We can put over 2.5 times as much charge on the Teflon filled capacitor as on the paper filled capacitor.

PROBLEM SG 16.7: Two large parallel plates in a vacuum chamber are separated by 1.00 mm. They are connected to a power supply of potential difference 2.5×10^3 V. A doubly charged oxygen ion (mass $= 2.8 \times 10^{-26}$ kg) starts from rest on the surface of one plate and is accelerated to the other. Neglecting gravity, what is the kinetic energy of the ion when it hits the other surface? Justify neglecting the force of gravity.

SOLUTION:

> **Procedure:** This problem can be solved in at least two ways. The easiest approach is to use conservation of energy, equating the initial potential energy to the final kinetic energy. The other approach is to use Newton's Laws directly to find the acceleration and then use the kinematical laws to get the kinetic energy.

The problem is illustrated in Figure SG 16.7. We assume the positive plate is at the bottom and that the ion is missing two electrons so it is repelled by the lower plate and attracted to the upper plate.

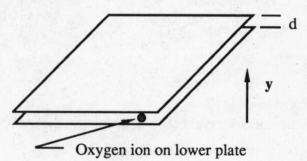

Oxygen ion on lower plate

Figure SG 16.7:   Parallel plate capacitor with a positive ion located on the positive plate.

FIRST METHOD:

We will use conservation of energy.

(Potential Energy + Kinetic Energy)$_{\text{Initial}}$ = (Potential Energy + Kinetic Energy)$_{\text{Final}}$

The initial kinetic energy is zero by assumption while the change in the potential energy is given by the product of the charge of the particle with the potential difference. If the particle charge is $2e$ where e is the electron charge and the potential difference is V, then

$$\text{(Kinetic Energy)}_{\text{Final}} = 2e\,V$$

$$= 2 \times 1.6 \times 10^{-19}\,\text{C} \times 2.5 \times 10^3\,\text{V} = 8.0 \times 10^{-16}\,\text{J}.$$

The final kinetic energy is 8.0×10^{-16} J.

We can justify neglecting the gravitational force by calculating the change in gravitational potential energy for the mass, m, moving a height $d = 1.00$ mm. The gravitational potential energy change is given by

$$m\,g\,d = 2.8 \times 10^{-26}\,\text{kg} \times 9.8\,\text{m/s}^2 \times 10^{-3}\,\text{m} = 2.7 \times 10^{-28}\,\text{J}.$$

This is clearly very much less than the significant figures for the kinetic energy produced by the electrical force, so we are justified in neglecting the effects of gravity.

SECOND METHOD:

Use the force laws to calculate the acceleration. Newton's third law gives

$$F = m\,a_y = q\,E_y$$

where a_y is the acceleration in the y-direction and E_y is the electric field. The electric field is given in terms of the potential difference by

$$E = \frac{V}{d}$$

so

$$a_y = \frac{2e}{m}\frac{V}{d}.$$

Since we don't know the time of travel between the plates, the obvious kinematical law to use is

$$v^2 = v_0^2 + 2\,a\,y.$$

Since the initial velocity is zero, this gives the velocity at the top plate as

$$v^2 = 2\,a\,d = \frac{4\,e}{m}\,V.$$

Hence the final kinetic energy is given by

$$\text{(Kinetic Energy)}_{\text{Final}} = \frac{1}{2}\,m\,v^2 = \frac{1}{2}\,m\frac{4\,e}{m}\,V = 2\,e\,V,$$

which is the same result as before.

Chapter 17
Electric Current and Resistance

Important Terms

Fill in the blanks with the appropriate word or words.

1. The SI system of current is the _____ which is a current flow of 1 Coulomb per second.

2. If the ratio of the applied voltage to the current flow through a circuit element is a constant, we say the circuit element obeys _____ Law.

3. A(n) _____ is a source of emf.

4. The _____ of a substance is the quantity which relates the resistance of a given object to its physical dimensions.

5. The _____ of a substance is the quantity which relates the electrical resistance of a given object to its physical dimensions.

6. When an unexpectedly large currents flow in a circuit, we say that we have a(n) _____ circuit.

7. The electric _____ is the rate at which charge passes through a conductor.

8. The SI unit of resistance is the _____.

9. _____ relates the electric power consumed to the circuit's voltage and current.

10. If the magnitude of the current in a circuit is a constant, we have a(n) _____ current.

11. If the conductors in a circuit are broken so that no current can flow, we say that we have a(n) _____ circuit.

12. The electrical potential that a circuit has with respect to the earth is called the _____ .

13. Electrical circuit elements which obey Ohm's Law are called _____ .

14. _____ rules govern the potential drops and currents in a circuit.

15. The ratio of the applied voltage to the resulting current is termed the _____ of the circuit.

Write the definitions of the following:

16. battery (17.1):

17. ampere (17.1):

18. electromotive force (emf) (17.1):

19. Kirchhoff's laws (17.6):

20. electric current (17.1):

21. Ohm's Law (17.2):

22. resistivity (17.3):

23. Joule's Law (17.4):

24. ohm (17.2):

25. direct current (dc): (17.1):

26. resistor (17.2):

27. short circuit (17.5):

28. open circuit (17.5):

29. resistance (17.2):

30. ground potential (17.5):

Answers to 1—15, Important Terms

1. ampere; 2. Ohm's; 3. battery; 4. electromotive force (or emf); 5. resistivity; 6. short; 7. current; 8. ohm; 9. Joule's Law; 10. direct; 11. open; 12. ground potential; 13. resistors; 14. Kirchhoff's; 15. resistance.

General Comments

For problems in this and successive chapters, we model real circuits with some simplified properties. For example, unless otherwise specified, we assume the wires connecting circuit elements are without resistance, capacitance, etc. This is an approximation to the real case where the wires have a much lower resistance than the circuit elements. We assume that voltage sources, such as a battery, are a constant voltage source independent of load unless we specifically indicate that the concepts given in section 17.9 are to be used; only then do we consider the internal resistance of a battery. We also neglect any edge effects in capacitors.

Nor is any consideration given in this chapter to how the resistivity of a metal and the insulation properties of non-metals arise. In Chapter 30 your authors discuss the fundamental physics that gives rise to resistivity. For this chapter we take the numbers in text Table 17.1 as experimentally measured numbers.

Sample Solutions

Problem SG 17.1: We are given a "black box" with two terminals. When we apply a known voltage to the terminals the current which flows is measured. This data is given in Figure SG 17.1. Graphically determine how the resistance of the "black box" depends upon voltage and plot it. Also plot the power dissipated in the box as a function of applied voltage. NOTE: The term "black box" means an object which we can not analyze by opening and taking it apart.

SOLUTION:

Procedure: Use the definition of resistance to make the graph. Use the definition of power in terms of voltage and current and make the plot.

The resistance is defined as the ratio of applied voltage to the current that flows. Read the current for each value of voltage and calculate the ratio. Check your results with those given in Table SG 17.1. Note that the resistance is defined as the ratio, not as the slope of the smooth curve.

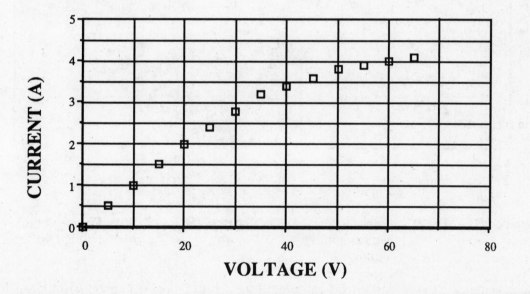

Figure SG 17.1: A plot of voltage versus current. You should draw a smooth curve through the data.

The power used by the circuit at each voltage is given by

$$\text{Power} = \text{Voltage} \times \text{current.}$$

These values are given in Table SG 17.1 also.

The plots requested are given in Figures SG 17.2 and 17.3

Table SG 17.1

applied voltage (Volts)	observed current (Amperes)	calculated resistance (Ohms)	Power (Watts)
0.0	0.0	0	0
5.0	0.5	10	2.5
10.0	1.0	10	10
15.0	1.5	10	23
20.0	2.0	10	40
25.0	2.4	10	60
30.0	2.8	11	84
35.0	3.2	11	1.1×10^2
40.0	3.4	12	1.4×10^2
45.0	3.6	12	1.6×10^2
50.0	3.8	13	1.9×10^2
55.0	3.9	14	2.0×10^2
60.0	4.0	15	2.4×10^2
65.0	4.1	16	2.6×10^2

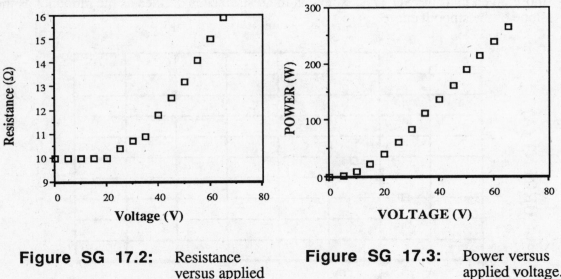

Figure SG 17.2:   Resistance versus applied voltage.

Figure SG 17.3:   Power versus applied voltage.

The resistance at 0 V cannot be calculated as a ratio; but, it can be calculated by determining the limit suggested by the values at V = 5 V, 10 V, and 15 V. We expect the correct value of this limit would be 10 Ω.

PROBLEM SG 17.2: It is desired to run an electric drill at the end of a 50 foot extension cord. The cord is made of 18 gauge copper wire (18 gauge wire has a cross-sectional area of 8.231×10^{-7} m^2). Under these conditions the drill is observed to draw 5.0 amperes. Assume the supply voltage is 100 V. What is the voltage at the drill? What would the voltage drop be if a saw drawing 25 Amperes was used on this cord?

SOLUTION:

> **Procedure:** Use the formula relating resistance, resistivity, length and cross-sectional area. Calculate the voltage drop for the given currents.

The voltage at the drill will be less than 100 V since there will be a voltage drop in the extension cord. Since we are given the current when the extension cord is used, we do not need to worry about the voltage-current behavior of the drill.

Text Table 17.1 can be used to get the resistivity of copper. The resistance of a length of copper is given by text Equation 17.4. Since the current flows down and back the wire, the total length of wire across which the voltage drop occurs is 100 ft.

$$100 \text{ ft} = 100 \text{ ft} \times \frac{12 \text{ in}}{1 \text{ ft}} \times \frac{0.0254 \text{ m}}{1 \text{ in}} = 30.48 \text{ m}.$$

So the total resistance in the wire (both directions) is given by

$$R = \frac{\rho L}{A} = \frac{1.72 \times 10^{-8} \ \Omega \text{ m} \times 30.48 \text{ m}}{8.231 \times 10^{-7} \text{ m}^2} = 0.637 \ \Omega,$$

where ρ is the resistivity of copper, l is the length of wire, and A is the cross sectional area of the wire.

The voltage drop when the drill is used is then given by

$$V = IR = 5.0 \text{ A} \times 0.637 \ \Omega = 3.2 \text{ V}.$$

When a 25 A saw is run the voltage drop will be 5 times larger or

$$V = 16 \text{ V}.$$

This is over 15% of the applied voltage, so the saw will be running at a much lower voltage than was probably built into the design of the saw by the manufacturer.

PROBLEM SG 17.3: How much does it cost to run the saw of Problem SG 17.2 for 5 hours? What is the cost of the energy wasted in heating the extension cord? Assume the cost of electricity is $0.09 per kilowatt-hour.

SOLUTION:

> **Procedure:** The total cost is given by: cost = Power times time(in hours) times cost per hour.

Since we know the current and the applied voltage we can find the power, so

$$\text{cost} = 100 \text{ V} \times 25 \text{ A} \times \frac{\text{kW}}{1000 \text{ W}} \times 5 \text{ hours} \times \$0.09 \text{ kilowatt/hr}$$

$$= \$1.13.$$

The power loss in the extension cord is given by the same formula where the voltage drop is 16 V. So, if we divide the $1.13 by 100 and multiply by 16, we should get the correct answer.

$$\text{cost of wasted power} = \frac{\text{cost} \times 16 \text{ V}}{100 \text{ V}} = \$ 0.18.$$

In terms of percentages, the loss is given by

$$\text{loss} = \frac{\text{waste}}{\text{comsumption}} \times 100 = \frac{\$.18}{\$1.13} \times 100 = 16\%.$$

About 16% of the power purchased is wasted in heating the extension cord.

PROBLEM SG 17.4: An automobile battery has a terminal voltage of 11.9 V when a current of 2.00 Amperes is drawn. When the battery is used to start the automobile a current of 75.0 A is drawn by the starting motor and the terminal voltage of the battery drops to 8.00 V. What is the emf and the internal resistance of the battery?

SOLUTION:

Procedure: Use the concept of internal resistance of a battery discussed in text section 17.9.

A battery can be considered as an ideal voltage source with a resistor in series as shown in Figure SG 17.4. The terminal voltage, V_T, is given by

$$V_T = R_L \times i_L = emf - R_I \times i_L,$$

where R_L is the external load resistance, R_I is the internal resistance of the battery, i_L is the current that flows from the battery when a resistance of R_L is attached across the battery, and *emf* is the true emf of the battery.

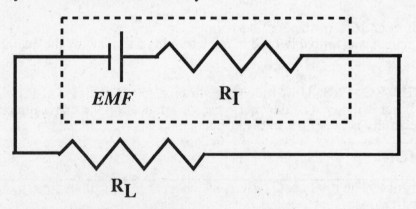

Figure SG 17.4:   Battery (dotted box) with resistor as load.

We are given the terminal voltage for two different current loads, so we should be able to find the emf and the internal resistance from

$$V_{\text{terminal}} = emf - I_{\text{load}} R_{\text{internal}} .$$

That is,

$$11.9 \text{ V} = emf - 2.00 \text{ A} \times R_{internal}$$

and

$$8.00 \text{ V} = emf - 75.0 \text{ A} \times R_{internal}.$$

To solve these, multiply the first equation by 37.5 and subtract the second equation, (See Appendix: Solving Simultaneous Equations) giving:

$$(446.25 - 8.00) \text{ V} = (37.5 - 1)\text{V} \times emf - (75 + 75)\text{A} \times R_{internal}$$

giving

$$438.25 \text{ V} = 36.5 \text{A} \times emf$$

or

$$emf = 12.0 \text{ V}.$$

Using this value for *emf* in both equations gives 0.050 Ω and 0.053 Ω respectively.

The process of subtracting the two 3 significant figure numbers will, in this case, produce a result that is known to only one significant figure, so we have

$$R_{internal} = 0.05 \text{ }\Omega.$$

The battery is an emf of 12.0 Volts with an internal resistance of 0.05 Ω

PROBLEM SG 17.5: A household circuit is wired for 115 V with a 20 Ampere circuit breaker. Which of the following items can be operated on that circuit at the same time? A 1800 watt toaster, a 2200 watt griddle, a 1000 watt iron, a 1500 watt mixer, lights totaling 400 watts. Assume the rules for direct current and power apply in this case.

SOLUTION:

Procedure: Use the definition of power in terms of voltage and current to determine the maximum wattage that the circuit breaker can support. The add wattages to determine what combination of light and appliances consume less power than the maximum allowed.

The easiest way to answer this question is to determine the maximum wattage the circuit can carry. Since power equals voltage times current, we have the information to calculate the total wattage the circuit can carry.

Let $V = 110$ V be the line voltage and $I_m = 20$ Amperes be the maximum current. Then neglecting any losses in the lines, the maximum power, P_m is given by

$$P_m = V \times I_m = 115 \text{ V} \times 20 \text{ A} = 2300 \text{ W}.$$

The total power consumption will be the sum of the individual power consumptions. First of all, we see that if the griddle is running, nothing else, not even the lights, can be running. For the other combinations, try adding wattages together to see if the sum is less than or equal to 2300 W. Any individual appliance and any appliance plus the lights may be on, but any two appliances consume more power together than the circuit breaker will take.

PROBLEM SG 17.6: What is the equivalent resistance of the combination of resistors shown in Figure SG 17.5?

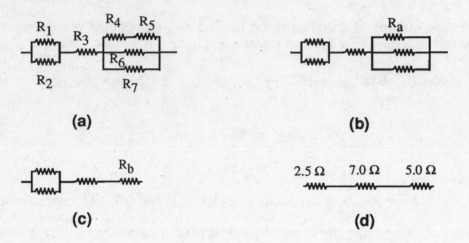

(a)   (b)

(c)   (d)

$$R_1 = R_2 = R_4 = R_5 = 5.0\,\Omega \quad R_3 = 7.0\,\Omega \quad R_6 = R_7 = 20.0\,\Omega$$

Figure SG 17.5: a). A combination of resistors.
b). The equivalent combination after the first reduction.
c). The equivalent combination after the second reduction.
d). The equivalent combination after the third reduction.

SOLUTION:

> **Procedure:** Use the rules for combining series and parallel resistors separately on parts of the circuit to simplify, repeating the process until you are left with only one resistor.

The first choice in simplifying resistor combinations is in deciding where to start. This choice is completely arbitrary; however, if you approach the combination systematically, you can minimize your work. The following procedure is best:

1. combine any resistors which are in series.
2. now combine resistors that are in parallel.

3. reexamine the combination repeating steps 1 and 2 until only one resistor is left. Consider the combination of seven resistors shown in Figure SG 17.5a. R_1 and R_2 are in parallel. R_4 and R_5 are in series and then they are in parallel with R_6 and R_7. These two combinations are in series with each other and with R_3.

The only resistors which are simply in series with each other are R_4 and R_5. Following rule 1, we combine these two resistors to get the equivalent resistance of

$$R_a = R_4 + R_5 = 5.0\,\Omega + 5.0\,\Omega = 10.0\,\Omega$$

as shown in Figure SG 17.5b.

We now should apply rule 2 to either of the two parallel combinations (right side or left side in Figure SG 17.5b). Since we started on the right side, the following will continue working on that set of resistors; however, you could proceed to reduce the left side first without changing the answer.

Applying rule 2, the right side parallel combination is reduced to

$$\frac{1}{R_b} = \frac{1}{10 \ \Omega} + \frac{1}{20 \ \Omega} + \frac{1}{20 \ \Omega} = \frac{4}{20 \ \Omega} = \frac{1}{5.0 \ \Omega} ,$$

which reduces the combination to that shown in Figure SG 17.5c.

We can still apply rule 2 to the combination of resistors on the left hand side of the combination giving

$$\frac{1}{R_c} = \frac{1}{R_1} + \frac{1}{R_2} = \frac{2}{5.0 \ \Omega} \quad \text{or} \quad R_c = 2.5 \ \Omega$$

as shown in Figure SG 17.5d.

Finally, apply rule 1 again giving us the equivalent resistance of $R_{eq} = 14.5 \ \Omega$.

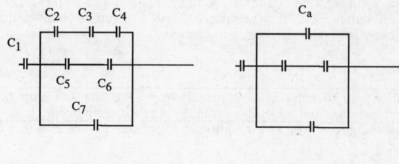

(a)

(b)

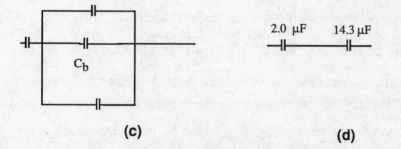

(c)

(d)

$$C_1 = C_5 = 2.0 \ \mu F \quad C_2 = C_3 = C_4 = 3.0 \ \mu F \quad C_6 = C_7 = 4.0 \ \mu F$$

Figure SG 17.6: a). The combination of capacitors.
b). The combination after the first reduction.
c). The combination after the second reduction.
d). The combination after the third reduction.

PROBLEM SG 17.7: Find the equivalent capacitance of the combination shown in Figure SG 17.6.

SOLUTION:

Procedure: Follow the rules described in Problem SG 17.6.

Examine the combination and note that the right hand side is a combination of capacitors in series and parallel. Since C_2, C_3, and C_4 are in series we combine them to get

$$\frac{1}{C_a} = \frac{1}{C_2} + \frac{1}{C_3} + \frac{1}{C_4} = \frac{1}{1.0\ \mu F}$$

as shown in Figure SG 17.6b .

Likewise, we combine C_5 and C_6 to get $C_b = 1.3\ \mu F$ as shown in Figure SG 17.6c.

Now, apply rule 2 to reduce the parallel combination to get $C_c = 6.3\ \mu F$ as shown in Figure SG 17.6d.

Finally combine the two series capacitors to get $C_{equivalent} = 1.6\ \mu F$.

PROBLEM SG 17.8: Use Kirchhoff's rules to find the current flowing through resistor R_5 in Problem SG 17.6 if the combination of resistors is connected across a 9.0 V battery. Neglect any internal resistance in the battery.

SOLUTION:

Procedure: Apply Kirchhoff's voltage rule to get the total current flowing and then apply the current rules and the voltage rules to get the current through R_5.

Since we have analyzed this combination in Problem 17.6 the majority of the work is done for us.

In general the following needs be done

1. Reduce the combination to the equivalent resistance.
2. Use the voltage rule to find the total current.
3. Use the voltage rule to find the voltage drops across major portions of the combination.
4. Use the current rule to calculate the current through a major subsection.
5. Repeat steps 3 to 4 as needed.

In particular we want the current through R_5, which by the current rule is the current through R_4 also. If we could determine the voltage drop across the sum of R_4 and R_5 we could calculate the current. This voltage drop is just the voltage drop across R_b in Figure SG 17.5c. With this in mind we apply the rules just given:

The equivalent resistance of the circuit was found in Problem SG 17.6 to be 14.5 Ω, so

$$i_{total} = \frac{V_{total}}{R_{eq}} = \frac{9.0\ V}{14.5\ \Omega} = 0.62\ A\ .$$

Since R_b is in series this must be the current through R_b. This allows us to calculate the voltage drop across R_b which we denote by V_b.

$$V_b = i_{total} R_b = 3.1 \text{ V}.$$

R_4 and R_5 together were denoted by the symbol R_a. The current through R_a and hence the current through R_5 is given by

$$i = \frac{V_b}{R_a} = \frac{3.1 \text{ V}}{10 \ \Omega} = 0.31 \text{ A}.$$

The current through R_5 is 0.31 A.

PROBLEM SG 17.9: What is the power consumption of an electric toaster if its resistance is 10.6 Ω and it operates on a household circuit?

SOLUTION:

> **Procedure:** Apply Ohm's Law.

This problem is similar to text Example 17.5. Here we are given the resistance (10.6 Ω) and the voltage (assume an effective household voltage of 120 V) and are asked to find the power. Using voltage and resistance we can easily find the current:

$$I = \frac{V}{R} = \frac{120 \text{ V}}{10.6 \ \Omega} = 11.3 \text{ A}.$$

We can then combine current and voltage to get the power.

$$P = IV = (11.3 \text{ A}) (120 \text{ V}) = 1360 \text{ W}.$$

The power consumption is 1360 Watts.

PROBLEM SG 17.10: A voltmeter of resistance 1000 Ω, placed in the circuit of Figure SG 17.7 as shown, reads a voltage of 40.0 V. Find the resistance R and the currents I, I_1, and I_2. What is the potential at point A when the voltmeter is not in the circuit?

SOLUTION:

> **Procedure:** First use Ohm's Law to calculate the currents I and I_2. Then use Kirchhoff's Laws to find I_1 and hence R. When we know R, we can calculate the voltage drop without the meter.

For this problem we only need to know that a voltmeter has a known resistance for a given measured voltage. Since the voltage drop across the voltmeter is 70 V, the voltage at A is:

$$V_A = 120.0 \text{ V} - 40.0 \text{ V} = 80.0 \text{ V}.$$

So there is a 80.0 V potential across the 500 Ω resistor and hence

$$I = \frac{V_A}{500\ \Omega} = 0.160\ \text{A}.$$

Likewise,

$$I_2 = \frac{\text{Voltmeter reading}}{\text{Voltmeter resistance}} = \frac{40.0\ \text{V}}{1000\ \Omega} = 0.040\ \text{A}.$$

Kirchhoff's Laws require that

$$I = I_1 + I_2,$$

so

$$I_1 = 0.160\ \text{A} - 0.040\ \text{A} = 0.120\ \text{A},$$

and

$$R = \frac{\text{Voltmeter reading}}{I_1} = \frac{40.0\ \text{V}}{0.120\ \text{A}} = 333\Omega.$$

The unknown resistance is 333 Ω, and the currents I, I_1, I_2 are 0.160 A, 0.120 A and 0.040 A respectively.

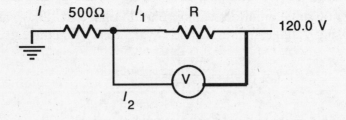

Figure SG 17.7: Circuit with unknown resistance R and voltage measured by a voltmeter of internal resistance 1000 Ω.

If the voltmeter were removed from the circuit, the total resistance of the circuit would be $R + 500\ \Omega$, so the current that flows would be:

$$I = \frac{120.0\ \text{V}}{R + 500\Omega} = 0.144\ \text{A},$$

and

$$V_A = I \times 500\ \Omega = 72.0\ \text{V}$$

So the true circuit voltage should be 72.0 V rather than the 80 V deduced with the voltmeter. That is to be expected since the voltmeter is in parallel and hence reduces the effective resistance between the source and point A, so a larger current flows and hence a larger voltage drop occurs across the 500 Ω resistor.

The voltmeter has too low a resistance to accurately measure this circuit. A better voltmeter will have a much higher internal resistance. Such a meter will read a voltage which is closer to the voltage that would be present in the meter were not attached.

Chapter 18
Magnetism

Important Terms

Fill in the blanks with the appropriate word or words. The three right hand rules are included as well as the important terms.

1. The magnetic dipole _____ is the value assigned to two opposite magnetic poles separated by a fixed distance.

2. A _____ uses a magnetic field to bend moving charged particles into a circular orbit.

3. The relationship among the size and direction of a small segment of current, the direction to a point and the magnetic field produced at that point is called the Law of _____ .

4. The _____ gives us the force that a magnetic field exerts on a current or a moving charge.

5. The _____ rule gives us the direction of the magnetic moment produced by a current loop.

6. A(n) _____ is a device for measuring currents.

7. The _____ rule for B gives us the magnetic field produced by a current or a moving charge.

8. _____ Law relates the average magnetic field around a closed path to the current enclosed by that path.

9. A(n) _____ is a device for measuring voltage.

10. A(n) _____ is a current measuring device which uses the magnetic field produced by the current being measured to deflect a magnetized compass needle.

11. The magnetic effects resulting from cooperative interactions of individual ionic moments is termed _____ .

12. The SI unit of magnetic field is the _____ , which has dimensions of newtons/ampere-meter.

13. _____ Law for magnetism states that the net magnetic flux through any closed surface is zero.

14. A(n) _____ is a helical winding that may carry current.

15. The _____ law governs the behavior of paramagnetism.

16. Regions of a ferromagnetic solid in which all the ionic moments are aligned in the same direction are called magnetic _____.

17. The _____ field is produced by an electric current in analogy to the electric charge producing an electric field.

18. The magnetic _____ relates the magnetism produced in a material to the externally applied magnetic field.

19. The _____ is the magnetic analogy of the electric dipole.

Write the definitions of the following:

20. magnetic domains (18.10):

21. right-hand rule for B due to a current (18.2):

22. tesla (18.3):

23. magnetic dipole (18.1):

24. cyclotron (18.5):

25. galvanometer (18.8):

26. magnetic moment (18.2):

27. Ampere's Law (18.9):

28. magnetic field (18.1):

29. Law of Biot and Savart (18.6):

30. Gauss's Law for magnetism (18.2):

31. ammeter (18.8):

32. voltmeter (18.8):

33. ferromagnetism (18.10):

34. solenoid (18.9):

35. right-hand rule for the magnetic moment of a current loop (18.7):

36. Curie law (18.10):

37. magnetic susceptibility (18.10):

38. right-hand rule for the force produced by a magnetic field acting on a current (18.2):

Answers to 1–19, Important Terms.

1..moment; 2. cyclotron; 3. Biot and Savart; 4. right-hand rule; 5. right-hand; 6. ammeter; 7. right-hand ; 8. Ampere's; 9. voltmeter; 10. galvanometer; 11. ferromagnetism; 12. tesla; 13. Gauss's; 14. solenoid; 15. Curie; 16. domains; 17. magnetic; 18. susceptibility; 19. magnetic dipole

General Comments

We model the effects that are called "magnetic" as we did for the effects called "electric". That is, there is an effect, called the magnetic,field. which is represented by the field lines. When scientists first introduced the magnetic field, the field lines were treated as real objects, rather than just representing the effects of the magnetic field.

The magnetic force laws are very similar to the electric force laws. The major difference is that whereas electric force laws contain cosines of angles, the magnetic force laws contain sines of angles. The magnetic force laws also involve directed sources (currents) and three dimensional relationships. For example the force on a current element in a magnetic field lies in a direction which is perpendicular to the plane defined by the current element and the magnetic field. The electric force, on the other hand, is always directed along the electric field.

To help keep the direction of magnetic forces straight a number of right-hand rules have been introduced. We use three of them in this chapter. The first is shown in J&C Figure 18.8, the second in J&C Figure 18.10, and the third in J&C Figure 18.20.

First right-hand rule: If the thumb of your right hand points the direction a current flows, your curled fingers point in the direction of the magnetic field produced by that current as shown in text Figure 18.8.

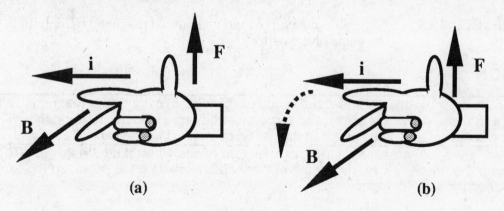

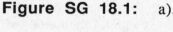

Figure SG 18.1: a). Identification of right-hand with current, magnetic field and force.

b). Right-hand screw rule.

Second right-hand rule: (right-hand screw rule): A current element and an external magnetic field define a plane (if they are not parallel). By the term "external" we mean do not consider the magnetic field produced by the current. The force the external field produces on the current element is perpendicular to the plane containing the current and the magnetic field and is directed the way a normal wood screw would advance if it were twisted the direction corresponding to a rotation of the current into the magnetic field through the smallest angle. To visualize this hold your right hand in front of you with your thumb upwards, your forefinger (or index finger) to the left and your next finger somewhat towards you as shown in Figure SG 18.1a. Let your index finger represent the current and your next finger the magnetic field, then your thumb points in the direction of the force. If you rotate your hand inward towards your body, making your index finger point towards the right and your next finger towards your body, then your thumb points in the direction that a normal (right-handed) screw would advance as shown in Figure SG 18.1b.

Third Right-Hand Rule: The magnetic dipole produced by a current loop can be determined by curling the fingers of your right hand in the direction of positive current in the loop. Your thumb points in the direction that the magnetic dipole points.

Sample Solutions

PROBLEM SG 18.1: (Exercises with the Second Right-Hand Rule) Suppose a current, i, and an external magnetic field, B, lie in the plane of the paper as shown in Figure SG 18.2a–c. What is the direction of the force on the current element in each case?

(a)

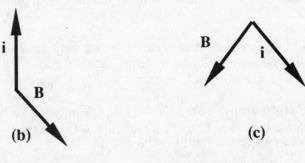

(b)   (c)

Figure SG 18.2: Examples of a current and an external magnetic field, both in the plane of the paper.

SOLUTION:

> **Procedure:** Apply the second right-hand rule. Since current and the magnetic field both lie in the plane of the paper, the force must be perpendicular to the paper. The question is then is the force up (out of the paper) or down (into the paper).

For case (a), if we arrange the first two fingers of the right hand along i and B as directed in Figure SG 18.1a, the thumb of our hand points into the paper, so the force is in a direction into the paper.

For case (b), although we may have sore fingers, the thumb points out of the paper.

Finally, for case (c), if we point the forefinger along i and the index finger along B, the thumb is again down.

When you do these exercises do not swing your fingers back and forth with respect to each other. In all cases keep the position of the thumb and fingers fixed as shown in Figure SG 18.1a. Arrange your hand by twisting it at the wrist and/or walk about the paper (or be lazy and move the paper around).

PROBLEM SG 18.2: A particle of electric charge q = 1.92×10^{-18} C is injected into a region with a magnetic field of 0.58 T with an initial velocity of 2.0×10^5 cm/s at an angle of 30° with respect to the magnetic field. What is the subsequent motion of the particle?

SOLUTION:

Procedure: Use the second right-hand rule and determine the magnitude of the force from Equation (18.2) of the text.

The motion and field described are sketched in Figure SG 18.3a. Coordinates have been chosen in which x $\Rightarrow$ y $\Rightarrow$ z is right handed. It is important that this sense be preserved if the right-hand rule is to be used. The magnetic field is indicated by *B* and the particle's motion by *v*. Only one line of magnetic field *B* is shown. *B* has the same value and direction at all points.

We will take the magnetic field as being directed along the y-axis, the particle's initial velocity to be in the x-y plane, and the 30° angle to the right of *B* in the x-y plane as shown. This was an arbitrary choice since *B* and *v* define a plane. It is simplest to let this plane be one of the coordinate planes. Our choice of the x-y plane is arbitrary.

As the problem is stated, we could have chosen *v* to be 30° to the left in the x-y plane instead of to the right. Our choice was arbitrary. The left hand choice will lead to a mirror solution. For the problem, as stated, either solution is acceptable.

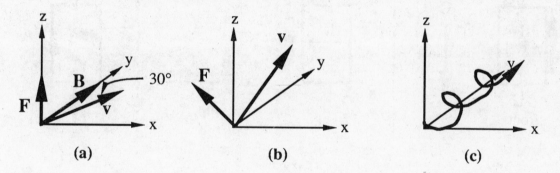

(a) (b) (c)

Figure SG 18.3: a). The initial velocity and force. Only one line is drawn to represent the uniform *B* in the y-direction.
b). The velocity and force at a later time.
c). The resulting motion.

Using the right-hand rule, we see that the initial force, *F*, on the particle will be exerted in the +z-direction and will be of magnitude

$$F = q v B \sin \theta$$

$$= 1.92 \times 10^{-18} \text{ C} \times 2.0 \times 10^3 \text{ m/s} \times 0.58 \text{ T} \times \sin(30°)$$

$$= 1.1 \times 10^{-15} \text{ N}.$$

The particle will be accelerated and hence gain velocity in the plane defined by z and the initial direction of *v*. The acceleration is a right angles to *v* (a right angle is a 90° angle), so it initially the velocity only changes the direction of *v*. Once the direction of *v* changes, the right-hand rule gives us a force that is still perpendicular to the new v as shown in Figure SG 18.3b. Thus, since the force is always perpendicular to v, the magnitude of v remains constant. This is the same arrangement that we had in Chapter 5 with centripetal force and uniform circular motion. The force is at right angles to the velocity and hence the force changes the direction of the velocity not its magnitude.

Thus the particle will move in a circular motion around the direction of B. Since the particle has some motion along the direction of B, the particle will spiral about the direction of B as shown in Figure SG 18.3c.

PROBLEM SG 18.3: A rectangular coil 1 cm wide by 2 cm long is suspended vertically at the top and bottom by wire fastened to the coil at the middle of the 1 cm sides as shown in Figure SG 18.4a. The coil contains 40 turns of wire. The two long sides each have a total mass of 20 g and the short sides each have a mass of 10 g. The coil is in a uniform magnetic field B which is initially perpendicular to the plane of the coil. When a current of 3.0 A flows through the coil it is deflected so that the plane of the coil is parallel to the magnetic field as shown in Figure SG 18.4b. If it takes 0.75 J of energy for this rotation, what is the magnitude of the field B? What torque is provided by the supporting wire in the final configuration?

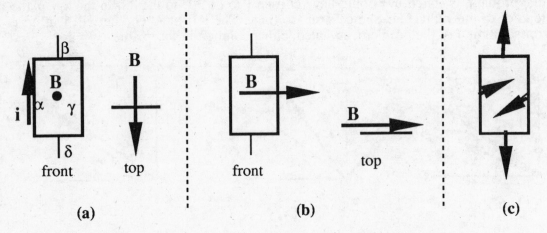

Figure SG 18.4: a). Initial configuration of coil and field.
b). Final configuration of coil and field.
c). Forces on the coil sides at an intermediate position.

SOLUTION:

> **Procedure:** Use text Equation 18.9 to calculate the magnetic moment of the loop. Then Equation 18.1 allows us to calculate the torque. By analogy the energy of the magnetic dipole is similar to Equation 15.4.

The magnitude of the force is given by text Equation 18.2 and its direction by the right-hand rule. The force acts on the current lengths involved. We have labeled the sides as α, β, γ, and δ. At all orientations between starting and stopping, the forces acting on sides β and δ tend to stretch the coil. The forces on sides α and γ try to rotate the coil as shown in Figure SG 18.4c or in J&C Figure 18.12. The torque is thus

$$\tau = N\,i\,A\,B \sin \theta$$

where

Quantity	Symbol	Value
number of turns	N	40
current	i	3.0 A
area of loop	A	2.0 cm²
magnetic field strength	B	unknown
angle between magnetic field and normal to plane of loop	θ	variable

When the rotation finishes $\theta = 90°$, while at the initial position $\theta = 0°$. The loop acts as a magnetic dipole with moment,

$$\mu = i\,N\,A = 2.4 \times 10^{-2}\ \text{A m}^2 .$$

Since the force equation for the magnetic dipole is the same as the electric dipole (see text Equation 15.4), the potential energy is given by

$$U = -\mu\,B\,\cos\theta.$$

The change in potential energy, ΔU, going from $\theta = 0°$ to $\theta = 90°$ is

$$\Delta U = -\mu B \cos(90°) - [-\mu B \cos(0°)] = \mu B.$$

By energy conservation

$$\Delta U = \text{work done} = 0.75\ \text{J} = 2.4 \times 10^{-2}\ \text{A m}^2\ B,$$

so

$$B = \frac{0.75\ \text{J}}{2.4 \times 10^{-2}\ \text{A m}^2} = 31\ \text{T}.$$

The torque applied by the support at maximum displacement must equal the magnetic torque since the system is in equilibrium. Thus the maximum torque provided by the wire is

$$\tau_{max} = \mu\,B = 0.75\ \text{N–m}.$$

PROBLEM SG 18.4: A long straight wire carries a current of 75 A. At what distance from the wire will the magnetic field strength of the magnetic field be 10 times the earth's field. Assume the earth's field is 5.0×10^{-5} T.

SOLUTION:

Procedure: Calculate the magnetic field produced by the wire and set that magnitude equal to ten times the magnitude of the earth's magnetic field.

The field produced by a long straight wire carrying a current i is given by text Equation 18.6 as

$$B. = k' \frac{2i}{d}$$

where B is the field and d is the perpendicular distance to the wire.

The direction of B is given by the right-hand rule. If we let B_E be the earth's magnetic field, the condition for the problem is

$$10\, B_E = 5.0 \times 10^{-4}\ \text{T} = k' \frac{2i}{d} = 10^{-7}\ N/A^2 \frac{150\ \text{A}}{d}$$

Hence,

$$d = 3.0 \times 10^{-2}\ \text{m} = 3.0\ \text{cm}$$

The field produced by the wire is equal to that of the earth's magnetic field at a distance of 3.0 cm.

PROBLEM SG 18.5: A cylindrical conductor of radius 0.10 cm carries a uniformly distributed current of 2.0 A. What is the magnetic field 0.050 cm from the center of the cylinder?

SOLUTION:

Procedure: Apply Ampere's Law. Since the wire is symmetric about the z-axis, only the current interior to the 0.050 cm distance will contribute to the magnetic field.

An end view of the cylindrical conductor is shown in Figure SG 18.5.

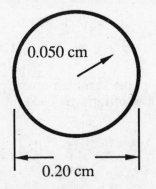

0.050 cm

0.20 cm

Figure SG 18.5: A cylindrical conductor of radius 0.10 cm carrying a uniform current.

The solution of this problem will involve an application of Ampere's Law. That is (see text Section 18.9)

$$\Sigma B\, \Delta l \cos\theta = \mu_0\, i$$

where the sum is taken over a loop surrounding the current i.

Since we are asked for the field at a radius of 0.05 cm from the axis without regard to the direction of this distance, we must choose the loop to be a circle of radius 0.050 cm with the circle's center at the center of the conductor.

That is, consider a circle of radius 0.050 cm centered on the z-axis and in the plane of Figure SG 18.5. Let this loop be approximated by a series of very short line segments of length Δl. Since the wire is symmetric about z, B will be the same for all segments.

Then $B \cos \theta$ gives us the field tangent to the circle and $\Sigma \Delta l$ gives us the circumference of the circle.

If we let d be the radius of the loop (0.050 cm) then

$$2\pi \, d \times B \; = \; (2\pi \times 5.0 \times 10^{-4} \text{ m}) \, B \; = \; \mu_0 \, i$$

where i is the current interior to 0.05 cm. We do not multiply out the 2π since μ_0 also will involve π.

The current is distributed uniformly across the wire, so

$$i \; = \; \frac{\text{area of circle of radius 0.05 cm}}{\text{area of circle of radius 0.10 cm}} \times (\text{total current})$$

$$= \; \frac{2\pi(0.050)^2}{2\pi(0.10)^2} \times 2.0 \text{ A}$$

$$= \; \frac{1}{4} \times 2.0 \text{ A} = 0.50 \text{ A}.$$

Hence

$$2\pi \times 5.0 \times 10^{-4} \text{ m} \times B \; = \; 4\pi \times 10^{-7} \text{ N/A}^2 \times 0.50 \text{ A}$$

so

$$B \; = \; 2.0 \times 10^{-4} \text{ N/A-m}$$

Note that the current further than 0.050 cm from the axis does not contribute to the magnetic field at 0.050 cm.

PROBLEM SG 18.6: A rectangular loop of wire carrying a current of 5 A is hung on the end of a spring in a uniform magnetic field of 0.010 T strength. The loop is 10.0 cm wide and 20.0 cm long and has a mass of 500 g. The magnetic field covers only part of the loop as shown in Figure SG 18.6. If the magnetic field (or the current) is turned off, the loop moves upward a distance of 10.0 cm. What is the spring constant?

SOLUTION:

Procedure: Use the magnetic force law to calculate the forces acting on each segment of the current. For equilibrium the sum of these forces must be equal and opposite to the force produced by the spring.

Since the loop moves upward when the magnetic field is off, the force on the loop must be as shown in the figure. Only F_1 produces any movement. Since the other forces are perpendicular to the spring, we can neglect them and any edge effects. The force F_1 is given by text Equation 18.1 as

$$F_1 = i l B \sin \theta$$

where B is the field strength, i the current, l the length of the current segment and θ is the angle between i and B ($\theta = 90°$ for this problem).

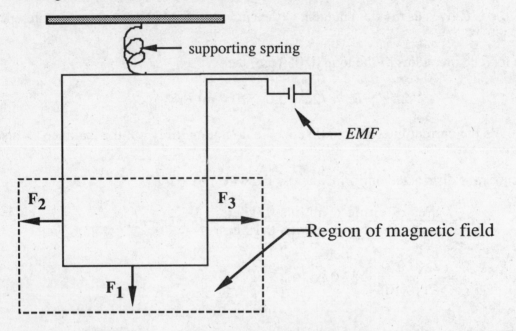

Figure SG 18.6:   Loop of wire hanging in a region of magnetic field. The magnetic field is perpendicular to the plane of the drawing. The magnetic forces are shown.

Thus

$$F_1 = 5.0 \, \text{A} \times 0.10 \, \text{m} \times 0.010 \, \text{T} = 5.0 \times 10^{-3} \, \text{N}.$$

This downward force must be matched by the upward force of the spring. That force would be given by $F = -kx$. Since the loop moves upwards 10 cm when the magnetic force is removed, the spring is distorted 10 cm from equilibrium. So

$$\Sigma \, \text{forces} = 0 = -k \times 0.10 \, \text{m} = 5.0 \times 10^{-3} \, \text{N},$$

or

$$k = 5.0 \times 10^{-2} \, \text{N/m}.$$

The mass of the loop does not enter into the solution since the position of the spring is measured from hanging equilibrium and since the gravitational force is constant in the problem.

Chapter 19
Electromagnetic Induction

Important Terms

Fill in the blanks with the appropriate word or words.

1. _____ Law states that the emf induced in a loop of wire is proportional to the rate of change of magnetic flux through the loop.

2. A current flowing through a coil sets up an induced current in the opposite direction. This effect is called _____ (or _____).

3. A current flowing through one coil causes a current to flow in another, nearby coil. This effect is called _____ inductance.

4. The unit of inductance in the SI system is the (volt-second/ampere) which is also called the _____.

5. A(n) _____ usually consists of two multiple turn coils of wire wound on the same iron core.

6. Oscillating electric charges generate _____ waves.

7. The emf produced by a changing magnetic field is called a(n) _____ emf.

8. A(n) _____ is any element of a circuit that exhibits inductance.

9. A(n) _____ converts a current unto rotational mechanical motion.

10. The current produced by a changing magnetic field is termed a(n) _____ current.

11. The _____ is a device which converts rotary motion into a direct voltage which can then produce a current.

12. A current which periodically reverses its direction is called a(n) _____ current.

13. _____ Law tells us that the current induced by a change in an external magnetic field is such that the magnetic field produced by the current will oppose the change in the external magnetic field.

14. The set of four equations that describe the behavior of electric and magnetic fields are called _____.

15. When an electric motor shaft begins to move, the motor produces a _____ in the opposite direction to the applied emf.

Write the definitions of the following:

16. electric motor (19.3):

17. alternating current (19.3):

18. self inductance (19.5):

19. henry (19.5)

20. Maxwell's equations (19.7):

21. transformer (19.4):

22. generator (19.3):

23. (mutual) inductance (19.5):

24. back (or induced) emf (19.3):

25. Faraday's law (19.1):

26. Lenz's Law (19.1):

27. electromagnetic waves (19.7):

28. induced emf (19.1):

29. inductor (19.5):

30. induced current (19.1):

Answers to 1–15, Important Terms.

1.. Faraday's; 2. self inductance (or just inductance); 3. mutual; 4. henry; 5. transformer; 6. electromagnetic; 7. induced; 8. inductor; 9. electric motor; 10. induced; 11. generator; 12. alternating; 13. Lenz's; 14.Maxwell's Equations; 15. back emf (or induced emf)

Sample Solutions

PROBLEM SG 19.1: An inductor is rated as being 1.2 H. What is the induced emf across its terminals if there is a current flowing which obeys the rule $i(t) = 2.00$ A $+ 0.50$ A/s $\times t$, where t is measured in seconds?

SOLUTION:

> **Procedure:** Use Faraday's Law to get the magnitude and the sign of the induced emf.

Faraday's Law (text Equation 18.2) gives us the relation between the inducted emf, E, and the changing current i

$$E = -L \frac{\Delta i}{\Delta t} = -1.2 \text{ H} \times 0.50 \text{ A/s} = -0.60 \text{ V.}$$

where L is the inductance and Δi is the current change that occurs in an elapsed time Δt.

Since the current is increasing, the induced emf must be negative to oppose the current change. This is shown in Figure SG 19.1 where the sign of the applied emf is indicated by the +,– pair on either side of the words "Applied emf" and the sign of the induced emf is indicated in the same manner.

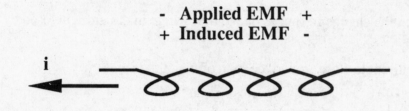

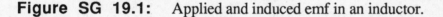

Figure SG 19.1:   Applied and induced emf in an inductor.

PROBLEM SG 19.2: The "U" shaped piece of wire is held fixed in a constant magnetic field of magnitude 0.80 T directed into the plane of the wire. The separation between the two parallel arms of the wire is 0.25 m. A conducting rod is moving away from the curve of the "U", as shown in Figure SG 19.2, at 3.0 m/s in contact with the wire. What current flows and what external force must be applied to the rod to maintain this velocity? Solve this problem first by assuming a total resistance for the complete circuit of 0.20 Ω and then indicate how the answer will change if the changing resistance of the circuit is included.

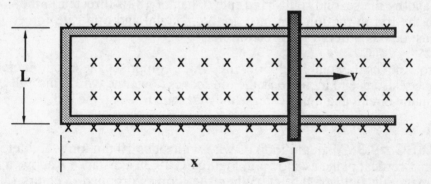

FIGURE 19.2:   A "U" shaped piece of wire with a bar sliding along the wire with a velocity v as shown. There is a uniform magnetic field B into the plane of the paper.

SOLUTION:

> **Procedure:** The rod across the ends of the "U" defines a circuit. As the rod moves the flux inside the circuit increases. Use Faraday's law to calculate the induced emf and then find the current that flows. The external magnetic field produces a force on this current which opposes the motion of the rod.

We let the symbol x represent the distance of the bar from the base of the loop, v the velocity of the rod and L the distance between the two parallel sides of the "U". Then from Faraday's law as given in text Equation 19.2,

$$E = -\frac{\Delta \Phi}{\Delta t} = -BL\frac{\Delta x}{\Delta t} = -BLv,$$

so

$$E = -0.80 \text{ T} \times 0.25 \text{ m} \times 3.0 \text{ m/s} = -0.60 \text{ V},$$

where we have taken the change in flux as due to the change in the area enclosed by the loop of wire and the bar.

This emf causes a current to flow around the loop of magnitude

$$i = \frac{E}{R} = \frac{0.60 \text{ V}}{0.20 \text{ } \Omega} = 3.0 \text{ A}.$$

The direction of the current can be determined in two steps. First, the magnetic field this current generates must oppose the external magnetic field and thus must point out of the paper. Then, the third right-hand rule (Chapter 18) tells us that this current must be counter-clockwise.

Text Equation 18.2 gives us the force exerted on the rod as

$$F = iLB = 0.6 \text{ N}$$

since the current and the rod are perpendicular. This force must oppose the motion of the rod and by the second right-hand rule (Chapter 18) is directed in the $-x$-direction. To keep the rod in motion we must apply an equal and opposite force. The applied force must be 0.6 N in the $+x$-direction.

As the rod moves the total length of wire which comprises the circuit increases, so we expect the resistance to increase as the rod moves down the wire. This will decrease the current in the circuit and thus decrease the magnetic force on the rod.

PROBLEM SG 19.3: A square loop of wire with sides 10 cm long and a total mass of 4 g is held in a vertical plane as shown in Figure SG 19.3a a distance h above a region with a uniform magnetic field of 0.50 T. When the loop is dropped it enters the region of magnetic field with velocity v as shown in Figure SG 19.3b. The loop has a resistance of 0.20 Ω. What must the distance h be so the velocity of the loop remains constant as it enters the region with the magnetic field?

SOLUTION:

Procedure: Calculate the rate of flux change in the coil, the resulting current, and hence the force on each side of the loop. For constant velocity this force must equal the gravitational force.

We let B be the magnetic field strength, L the length of a side of the loop, R the resistance of the loop, and h the initial height of the loop above the region of the magnetic field. When the loop is immersed y units into the region of magnetic field the flux inside the loop is given by (see Figure SG 19.3b):

$$\text{flux} = \Phi = B L y$$

where y is the distance the loop is immersed in the magnetic field region.

The change in flux is due solely to the change in y so the emf is given by

$$\text{emf} = -\frac{\Delta\Phi}{\Delta t} = -B L \frac{\Delta y}{\Delta t} = -B L v,$$

and the current that flows has a direction given by the Faraday's law and a magnitude given by

$$\text{current} = \frac{\text{emf}}{R} = \frac{B L v}{R}.$$

The external magnetic field acts on this current producing a force given by J&C Equation 18.2. For the horizontal side this force is

$$\text{Force} = B L \times \text{current} = \frac{B^2 L^2 v}{R}$$

and in a direction so as to oppose the motion of the wire as shown in Figure SG 19.3c. This figure also shows the forces on the vertical sides. Their magnitudes are equal and given by

$$\text{Force on vertical sides} = B I y.$$

These latter forces are equal and opposite so they do not affect the motion of the loop.

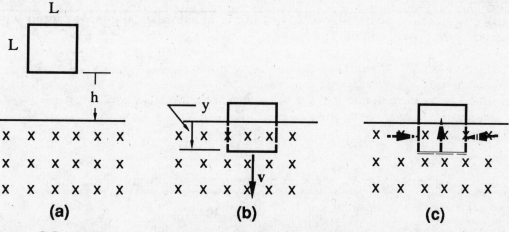

Figure SG 19.3: a). A square loop at rest a distance h above a region containing a uniform magnetic field. The magnetic field is into the plane of the paper.

b). The same loop having fallen into the magnetic field a distance y and having a velocity v.

c). The forces due to the current set up by the magnetic field. These forces act on the three sides of the loop that have entered the region of magnetic field.

The velocity when the loop first enters the region of magnetic field is the velocity when the loop has fallen a distance h. Equating the initial potential energy to the final kinetic energy gives us

$$mgh = \frac{1}{2} m v^2,$$

thus

$$v = \sqrt{2gh}.$$

For constant velocity the magnetic force is equal and opposite to the gravitational force so

$$mg = \frac{B^2 L^2 v}{R} = \frac{B^2 L^2 \sqrt{2gh}}{R}.$$

Solving this for h gives

$$h = \frac{1}{2} \frac{gm^2 R^2}{(BL)^4} = 0.50 \text{ m}.$$

The loop must be dropped from a height of 50 cm.

PROBLEM SG 19.4: A transformer attached to a 10 kV transmission line has 20 times as many primary turns as it does secondary turns. If the output line draws 50 A, what is the current draw on the input line?

SOLUTION:

> **Procedure:** Assume conservation of energy; that is, the power input is the same as the power output.

If we neglect the possibility of losses in the transformer, the power input will be the same as the power output. Thus

$$\text{power}_{\text{input}} = I_{\text{in}} V_{\text{in}} = \text{power}_{\text{output}} = I_{\text{out}} V_{\text{out}}$$

where I is the current and V is the voltage. Solving for I_{in} gives

$$I_{\text{in}} = \frac{V_{\text{out}}}{V_{\text{in}}} I_{\text{out}}.$$

For a transformer with N_{in} turns on the input and N_{out} turns on the output,

$$V_{\text{out}} N_{\text{in}} = V_{\text{in}} N_{\text{out}},$$

so

$$\frac{V_{\text{out}}}{V_{\text{in}}} = \frac{N_{\text{out}}}{N_{\text{in}}}.$$

Combining these gives

$$I_{\text{in}} = \frac{N_{\text{out}}}{N_{\text{in}}} I_{\text{out}} = 2.5 \text{ A}.$$

The primary current draw is 2.5 A.

PROBLEM SG 18.5: What is the equivalent inductance for the circuit shown in Figure SG 18.4a? Neglect any mutual inductance and assume that inductors in series and parallel add just as resistors in series and parallel add..

SOLUTION:

Procedure: Reduce the circuit to simpler components one step at a time as was done with resistor and capacitor networks in Chapter 17.

Problems such as this with inductors or others with capacitors are best approached in steps as was done with the resistor networks of Chapter 17. First take any simple series combinations and reduce them to a single inductor, then combine simple parallel elements. Repeat these two steps until only one circuit element remains. Starting with the circuit in Figure SG 18.4a this reduction is shown step by step in Figures SG 18.4b to 18.4d.

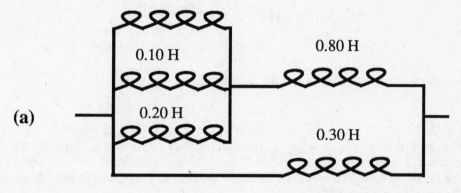

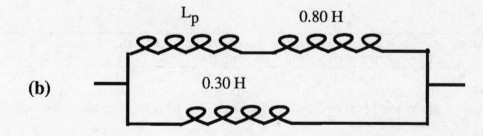

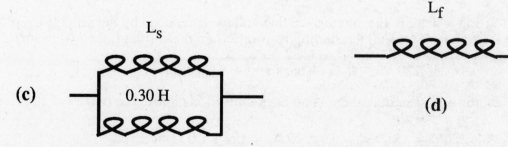

FIGURE SG 19.4: a). Series and parallel combination of inductors.
b). First reduction;
c). Second reduction; and
d). Final equivalent inductance.

Since there are no simple series inductors in this circuit, the first step is to combine the three inductors (0.10 H, 0.20 H and 0.50 H) which are in parallel. Using the same rules that apply to resistors:

$$\frac{1}{L_p} = \frac{1}{0.10 \text{ H}} + \frac{1}{0.20 \text{ H}} + \frac{1}{0.50 \text{ H}} = 17/\text{H},$$

so

$$L_p = 0.059 \text{ H} \qquad \text{(see Figure SG 18.4b)}$$

We now can combine L_p with the 0.80 H inductor to get

$$L_s = 0.059\text{H} + 0.80 \text{ H} = 0.86 \text{ H} \qquad \text{(see Figure SG 18.4c)}$$

Finally,

$$\frac{1}{L_f} = \frac{1}{0.86 \text{ H}} + \frac{1}{0.30 \text{ H}} = 4.50/\text{H}$$

so

$$L_f = 0.22 \text{ H.}$$

The equivalent inductance of this circuit is 0.22 H.

PROBLEM SG 19.6: A motor which operates on 220 V has an internal resistance of 1.0 Ω. If the motor draws 5.0 A when running, what is the back emf produced by the motor? If the maximum current that the motor can draw without damage to its coils is 20 A, what resistance must be added in series to limit the current to 20 A maximum? Such a series resistor is called a starting resistor.

SOLUTION:

> **Procedure:** The back emf must reduce the effective emf to that produced by 5.0 A through 1.0 Ω. The starting resistor must be such that 220 V can only provide 20 A.

When the motor is running the moving coils cut the magnetic field inside the motor and: therefore, must produce a back (or induced) emf. It is the vector sum (hence the difference) of the applied emf and the induced emf that provides the effective emf. That is, across the motor terminals:

$$\text{applied emf} - \text{back emf} = \text{resistance} \times \text{current.}$$

With the starting resistor in the curcuit, the line voltage is reduced by the voltage drop across the starting resistor. Call the starting resistor R_S and the current I, then we want

$$220 \text{ V} = R_S \times I + \text{back emf} + R_{\text{motor}} \times I.$$

When the motor is first started, there is no back emf. R_S is choosen to make $I = 20$ A. Thus

$$220\text{V} = (R_S + R_{\text{motor}}) \times 20 \text{ A} = (R_S + 1.0 \ \Omega) \times 20 \text{ A},$$

and hence

$$R_S = 10 \ \Omega$$

when running $I = 5$ A, so

$$\text{back emf} = 220 \text{ V} - 10 \ \Omega \times 5 \text{ A} - 1.0 \ \Omega \times 5 \text{ A} = 165 \text{ V.}$$

Chapter 20
Alternating-Current Circuits

Important Terms

Fill in the blanks with the appropriate word or words.

1. The time it takes the current in an inductor to build up to 1/e of its final value is called the inductive _____.

2. A(n) _____ (or a(n) _____) passes current in one direction and blocks the current in the other direction.

3. The capacitance factor analogous to resistance is called the capacitive _____.

4. The characteristic time for the current in an RC circuit to decay to 1/e of its initial value, when the EMF is removed, is called the _____.

5. The process of putting some of the output of an amplifier into the input is called _____.

6. The peak AC voltage is related to the peak AC current by the _____.

7. The constant of proportionality between the current and applied voltage for an inductor is called the inductive _____.

8. A(n) _____ amplifier is one of a large class of amplifiers whose behavior can be adjusted by the appropriate choice of feedback.

9. The _____ voltage is the square root of the of the cycle average of the voltage squared.

10. A(n) _____ is a two dimensional quantity which rotates at constant frequency. It is used to represent an oscillating current or voltage.

Write the definitions of the following:

11. inductive time constant (20.1):

12. capacitive time constant (20.2):

13. rms current and voltage (20.3):

14. operational amplifier (20.8):

15. phasor (20.5):

16. feedback (20.8):

17. capacitive reactance (20.4):

18. inductive reactance (20.4):

19. impedance (20.5):

20. rectifier (20.7):

21. diode (20.7):

Answers to 1–10, Important Terms

1. time constant; 2. diode, rectifier; 3. reactance; 4. capacitive time constant; 5. feedback; 6. impedance; 7. reactance; 8. operational; 9. rms (root-mean-square); 10. phasor

General Comments

Our model for this and the previous chapter is to assume that we can have perfect inductors, capacitors, and resistors. For example, an inductor is just an inductor without capacitance and resistance. All real inductors will have some capacitance and resistance, all real capacitors will have some resistance and inductance and all real resistors will have some inductance and capacitance. A properly designed circuit element will minimize these often unwanted effects. However, at high frequencies and large currents the real nature of the circuit elements must be taken into consideration when circuits are designed. Unless otherwise indicated in the problem, you should not need to consider the real nature of the circuit elements.

Sample Solutions

PROBLEM SG 20.1: The switch is closed in the circuit shown in Figure SG 20.1 at 11:45:00.0 AM. At what time will the current be at 90% of its maximum value?

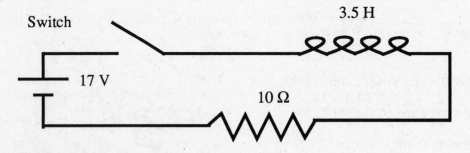

Figure SG 20.1:   Series LR circuit.

SOLUTION:

> **Procedure:** We are asked for the time dependence of the current which is given by text Equation 20.1. Since there is a dc voltage, the maximum current is as if the inductor were not present.

If we let t be the time since the switch was closed, V be the voltage, R the resistance and L the inductance then the current, i, is given by

$$i(t) = \frac{V}{R}\left(1 - e^{-\frac{Rt}{L}} \right) = i_m\left(1 - e^{-\frac{Rt}{L}} \right).$$

After a very long time the current will approach its maximum value i_m, given by

$$i_m = \frac{V}{R}.$$

We are asked for the time at which $i(t)/i_m = 0.90$, or the time t for which

$$0.90 = 1 - e^{-\frac{Rt}{L}}.$$

Thus

$$e^{-\frac{Rt}{L}} = 0.10.$$

The inverse of the exponential function was discussed in Chapter 13 and is denoted by *ln*. Most scientific calculators have both a *ln* and a *log* key; use the *ln* key since the *log* key is the inverse of 10^x and we want the inverse of e^x. Hence

$$-\frac{Rt}{L} = \ln(0.10),$$

or

$$t = -\frac{L \ln(0.10)}{R} = -\frac{3.5 \text{ H} \times (-2.3)}{10 \, \Omega} = 0.81 \text{ s}.$$

The current is at 90% of its maximum value 0.81 s after the switch is closed. At that time the clock reads 11:45:00.8 s.

PROBLEM SG 20.2: An inductor is attached to a voltage source whose peak amplitude can be kept constant but whose frequency can be varied. Keeping the voltage peak amplitude constant at 50 V while varying the frequency from 0.00 cycles per second (cps) to 100 cps produces the curve of reciprocal peak current versus frequency shown in Figure SG 20.2. The curve has a value of 0.10/A at frequency $f = 0.00$ and 126/A at $f = 100$ cps. At any frequency above 1 cps the curve appears to be a straight line. What is your best guess as to the ideal circuit elements which can approximate this real inductor in the frequency range of 0–100 cps?

SOLUTION:

Procedure: See if the given properties can be satisfied by a series RLC circuit. If not it will be necessary to assume components in parallel.

The peak voltage and current are related by $V_{peak} = Z\, i_{peak}$ where, if we assume the real component can be replaced by a series combination of resistor, inductor and capacitor, Z is given by text Equation 20.17 as

$$Z = \sqrt{R^2 + \left(2\pi f L - \frac{1}{2\pi f C} \right)^2}.$$

So

$$\frac{1}{i_{peak}} = \frac{Z}{V_{peak}}.$$

At very low frequencies

$$Z \approx \frac{1}{2\pi f C}$$

which becomes very large when $f \ll 1/C$. Since the reciprocal of the peak current is small at $f = 0$, we can have no series capacitance. Another way to state this is to take the series capacitance as infinite. An infinite series capacitor has no effect on a circuit since there can be no voltage across it (voltage = charge divided by capacitance). Remember the formula for a parallel plate capacitor from Chapter 16 had the capacitance proportional to the reciprocal of the plate separation. If the plate separation goes to zero the capacitance goes infinite and the capacitor becomes a short circuit.

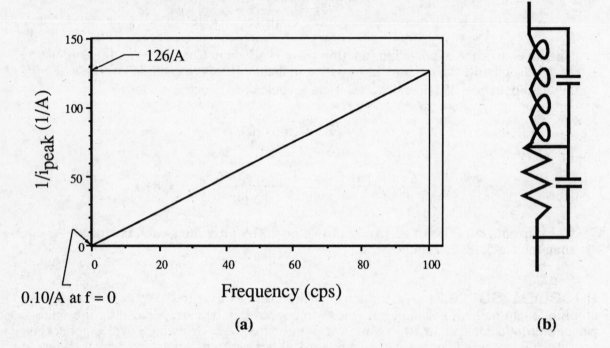

(a)   (b)

Figure SG 20.2: Reciprocal of peak current versus frequency with constant peak voltage for a real inductor.

The term with C should be left out of the expression for impedance and hence

$$\text{at } f = 0.00, Z = R$$

and

$$\frac{1}{i_{peak}} = \frac{R}{V} = \frac{0.1}{A} = \frac{R}{50 \text{ V}},$$

so

$$R = 5 \, \Omega.$$

Since we are told that the curve is essentially a straight line for frequencies above 1 cps, it will be reasonable to neglect R in the expression for Z giving

$$Z = 2\pi f L$$

and

$$\frac{1}{i_{peak}} = \frac{Z}{V} = 0.04 \, \pi f L.$$

The slope of the curve is

$$slope = \frac{\Delta\left(\frac{1}{i_{peak}}\right)}{\Delta f} = 0.04 \, \pi L = \frac{126/A - 0}{100 \text{ cps} - 0} = 1.26 \frac{1}{\text{A-cps}},$$

so

$$L = \frac{slope}{0.04\pi} = 10 \text{ H.}$$

Thus the inductor is a 10 H inductor which has an effective series resistance of 5 Ω. This is not a unique analysis of the circuit since there could be, and probably is, an effective parallel capacitance as shown in Figure SG 20.2b, however, since the curve is essentially a straight line except near $f = 0$, any such capacitance is too small to be detected in the 0-100 cps range.

PROBLEM SG 20.3: Given an inductor of 0.010 H, what value of capacitance would you need in order to build a series circuit that resonates at 50,000 Hz?

SOLUTION:

Procedure: The circuit described is an RLC circuit but we are not given the value of the resistance. Since we are not asked to find the impedance at resonance, the resistance value is not needed.

The resonant frequency of a series RLC circuit is given by

$$f_0 = \frac{1}{2\pi\sqrt{LC}}$$

so given f_0, the resonant frequency, and L, the inductance, C must be given by

$$C = \frac{1}{4\pi^2 f_0^2 L} = \frac{1}{39.48 \times 25 \times 10^8/s^2 \times 0.010 \text{ H}} = 1.0 \times 10^{-9} \text{ F.}$$

A 0.001 μF capacitor will be needed to make as series resonance with a 0.010 H inductor.

PROBLEM SG 20.4: A series RLC circuit has the behavior shown in Figure SG 20.3a at low frequencies and the behavior shown in Figure SG 20.3b at high frequencies. In both cases the lines are nearly linear (straight). In both cases the current is the RMS current and the applied RMS voltage is 100 V. Neglecting the series resistance, what is the predicted resonant frequency of this series RLC circuit?

SOLUTION:

> **Procedure:** Analyze the low and high frequency behavior of an RLC impedance neglecting the resistance of the circuit. At low frequencies the capacitance dominates the impedance while at high frequencies the inductance will dominate.

The necessary analysis is similar to that of Problem SG 20.4. Since $V_{rms} = Z\, i_{rms}$ with Z given by text Equation 20.20, we need to analyze Z at high and low frequencies. The terms high and low make sense by meaning capacitive reactance (low frequency) or inductive reactance (high frequency).

For low frequency

$$Z = \frac{1}{2\pi f C}$$

so

$$i_{rms} = 2\pi C V_{rms} f.$$

Thus i_{rms} versus f is a straight line, through $i_{rms} = 0$ at $f = 0$, whose slope is given by $2\pi C V_{rms}$.

Measuring the graph in Figure SG 20.5a gives us

$$2\pi C V = 628\, C = \text{slope} = \frac{6.28 \times 10^{-4}\,\text{A} - 0\,\text{A}}{100\,\text{cps} - 0\,\text{cps}},$$

$$C = 1.0 \times 10^{-8}\,\text{F} = 0.01\,\mu\text{F}.$$

For high frequency

$$Z = 2\pi f L$$

so

$$\frac{1}{i_{rms}} = \frac{2\pi L}{V_{rms}}\, f.$$

Thus $1/i_{rms}$ versus f is a straight line whose slope is $2\pi L/V_{rms}$. Measuring the graph in Figure SG 20.5b gives us

$$2\pi L/V_{rms} = 6.28 \times 10^{-2}\,L = \text{slope} = \frac{62.9 - 0.1}{2.0 \times 10^4 - 1.0 \times 10^4}\,/\text{A-s}.$$

so

$$L = 0.10\,\text{H}.$$

The inductance is 0.10 H and the capacitance is 0.01 μF. This gives a resonant frequency f_0 of

$$f_0 = \frac{1}{2\pi\sqrt{LC}} = 5{,}000\,\text{cps}.$$

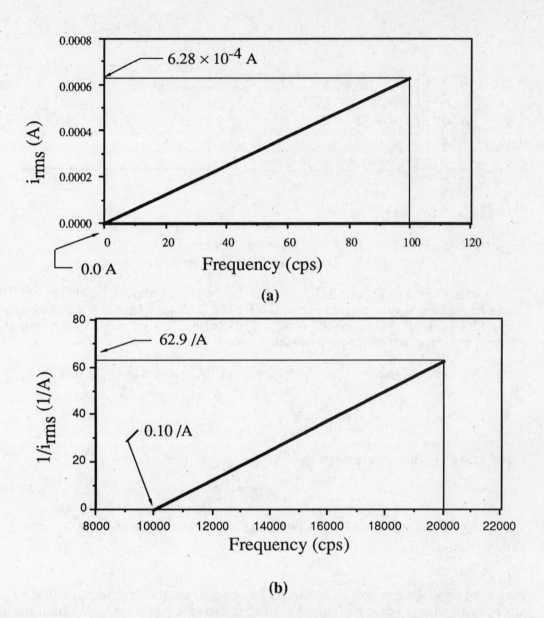

Figure SG 20.3:   a). Low frequency behavior of a series LC circuit.
  b). High frequency behavior of a series LC circuit.

PROBLEM SG 20.5: The transformer in a CD audio system has an rms output of 12.4 V. What is the rms voltage across the load resistor R shown in Figure SG 20.4a?

SOLUTION:

Procedure: The diode cuts off any negative voltage, so the output will be a series of peaks separated by periods of no voltage as shown in text Figure 20.20. These periods of no voltage will affect the rms value obtained from the definition of rms. The rms input voltage will enable you to find the peak output voltage. Then apply the definition of rms again to answer the question.

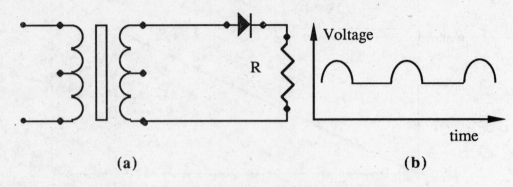

Figure SG 20.4:   a). A half wave rectifier circuit.
b). The output wave form for a sine wave input.

The circuit shown in Figure SG 20.4a has an output as shown in Figure SG 20.4b. These figures are similar to text Figure 20.20. We let V_{rms} be the input rms voltage, V_0 the input peak voltage, V_r be the peak output voltage above zero, V_a the output rms voltage, and v the instantaneous voltage.

Text Equation 20.10 allows us to find the size of V_0 in terms of V_{rms}, that is

$$V_{rms} = \sqrt{\overline{v^2}} = \frac{V_0}{\sqrt{2}}$$

gives, using the first and last terms,

$$V_0 = \sqrt{2}\, V_{rms}.$$

We assume that there is no loss in the diode so that the value of V_r is the same as V_0, that is

$$V_r = V_0 = \sqrt{2}\, V_{rms}.$$

However when we use text Equation 20.10 again to get the rms output voltage V_a, the answer will change since the output voltage is zero for half a cycle. Without the diode we would have had the same contribution form the negative portion of the voltage as from the positive portion. With the diode the negative portion is zero.

Without the diode

$$\overline{v^2} = \frac{V_0^2}{2}, \qquad \text{(without diode)}$$

while with the diode we must get only half of that value, that is

$$\overline{v^2} = \frac{V_r^2}{4}. \qquad \text{(with diode)}$$

Thus

$$V_a = \sqrt{\overline{v^2}} = \sqrt{\frac{V_r^2}{4}} = \frac{V_r}{2}.$$

For our particular problem $V_{rms} = 12.4$ V, so $V_r = 17.5$ V and hence $V_a = 8.77$ V.

The rms output voltage is 8.77 V.

PROBLEM 20.6: What is the total gain of the two operational amplifier combination shown in Figure SG 20.5 if $R_1 = 2$ kΩ, $R_2 = 10$ kΩ, $R_3 = 5$ kΩ, and $R_4 = 25$ kΩ.

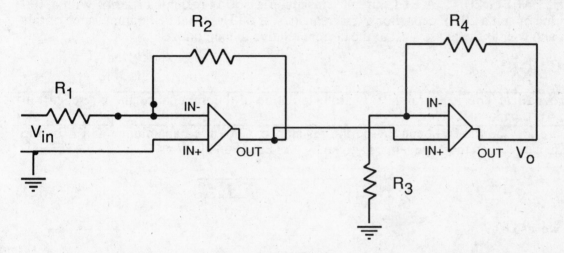

Figure SG 20.5: An inverting amplifier and a non-inverting amplifier connected in series. On the figure IN+ is the non-inverting input, IN– is the inverting input, and OUT indicates the output of the amplifier. If wires cross without a heavy dot, there is no connection.

SOLUTION:

Procedure: Treat this problem as two amplifiers. The output of the first amplifier is the input for the second amplifier. Since the voltage output is the gain times the voltage in for each amplifier, we expect the total gain to be the product of the gains of each amplifier

Let V_t be the output voltage of the first amplifier, G_1 be the gain of the inverting stage, G_2 be the gain of the non-inverting stage, V_{in} be the input voltage, and V_0 be the final output voltage. Referring to Figure SG 20.5 we can express the gains of each stage in terms of the resistor values as:

$$G_1 = -\frac{R_2}{R_1},$$

and

$$G_2 = \frac{R_3 + R_4}{R_3}.$$

Then V_t is given by

$$V_t = G_1 V_{in}$$

and

$$V_0 = G_2 V_t = G_2 G_1 V_0 = GV_{in} .$$

So the total gain, G, is given by

$$G = G_2 G_1 = -30.$$

The output voltage is inverted and 30 times the input voltage.

PROBLEM SG 20.7: A RCL circuit consists of a 500 Ω resistor in series with a 100 mH inductor and a 20 μF capacitor. At a frequency of 60 Hz, what is the impedance of this circuit and what is the phase shift of the current relative to the voltage?

SOLUTION:

Procedure: This is a straight forward calculation of the impedance and phase shift in an AC circuit.

If we let f be the frequency R be the resistance, C be the capacitance and L be the inductance, then the impedance is given by text Equation 20.17 as

$$Z = \sqrt{R^2 + \left(2\pi fL - \frac{1}{2\pi fC}\right)^2} .$$

If we also let

$$X_C = \frac{1}{2\pi fC}$$

and

$$X_L = 2\pi fL ,$$

then the phase shift is given by

$$\phi = \tan^{-1}\left(\frac{X_L - X_C}{R}\right)$$

and we can also write:

$$Z = \sqrt{R^2 + (X_L - X_C)^2} .$$

A substitution of the numbers will give the requested result. We find

$$X_C = 132 \ \Omega, X_L = 37.7 \ \Omega, \phi = -11° \text{ and } Z = 509 \ \Omega.$$

Chapter 21
Geometrical Optics

Important Terms

Fill in the blanks with the appropriate word or words. The laws of reflection and refraction as well as the thin lens equation are also included in these exercises.

1. The straight lines used to represent the path of light is called a light _____ .

2. The angle between the incoming light and the normal to a surface is called the _____ .

3. The law of refraction is also known as _____ law.

4. If the angle of incidence is greater than the critical angle, _____ will occur.

5. A lens whose thickness is small compared with its focal length is called a(n) _____ lens.

6. When light rays behave as if they are diverging from a point without actually ever having been at that a point we say that a(n) _____ image has been formed.

7. The _____ of a lens or mirror are the failures of that lens or mirror to give a perfect image.

8. _____ light has frequencies just below (lower) those of visible light.

9. The plane of _____ contains the normal, the incident ray, and the reflected ray.

10. The _____ is the minimum angle of reflection at which all internal light is totally reflected.

11. A lens that bends light so as to separate parallel rays away from each other is called a(n) _____ lens.

12. When the light rays that form an image actually intersect at that image, we have a(n) _____ image.

13. The _____ magnification of an image is the ratio of the image size to the size of the object that produced that image.

14. The spreading of light by wavelength as the light passes through a substance is the _____ of that substance.

15. _____ light has frequencies just above (higher) those of visible light.

16. The law of _____ gives us the relation between the angle of incidence of a light ray and the angle at which that ray is reflected.

17. The _____ for a substance is the ratio of the speed of light in a vacuum to the speed in the substance.

18. A lens that bends light so as to bring parallel light rays towards each other is called a(n) _____ lens.

19. The _____ of a thin converging lens is the distance from the lens that initially parallel rays finally meet.

20. The thin lens _____ relates object and image distances to the focal length of the lens.

21. The angle between a reflected light ray and the normal to the surface that reflected that ray is called the angle of _____.

Write the definitions of the following:

22. dispersion (22.8):

23. ultraviolet (22.1):

24. total internal reflection (21.3):

25. aberrations (22.8):

26. infrared (21.1):

27. converging lens (21.4):

28. critical angle (21.3):

29. lateral magnification (21.9):

30. index of refraction (21.2):

31. angle of incidence (21.2):

32. thin lens equation (21.6):

33. diverging lens (21.4):

34. Snell's Law (21.2):

35. virtual image (21.5):

36. angle of reflection (21.2):

37. plane of incidence (21.2):

38. real image (21.5):

39. thin lens (21.4):

40. law of reflection (21.2):

41. focal length (21.4):

42. ray (21.1):

Answers to 1-21, Important Terms

1. ray; 2. angle of incidence; 3. Snell's; 4. total internal reflection; 5. thin; 6. virtual;
7. aberrations; 8. Infrared; 9. incidence; 10. critical angle; 11. diverging; 12. real;
13. lateral (or linear); 14. dispersion; 15. Ultraviolet; 16. reflection; 17. index of refraction;
18. converging; 19. focal length; 20. equation; 21. reflection

Comments on Models

In this chapter we use the idea of light rays to represent the travel of light. The light ray is
the normal to the wavefront. Don't confuse the light rays with the light itself. The rays only
represent the direction the light is traveling.

Sample Solutions

PROBLEM SG 21.1: A mirror tilted 45° to the horizontal faces another mirror which
is tilted 30° to horizontal as shown in Figure SG 21.1a. Light falls vertically downward
onto the first mirror as shown in the figure. It is reflected from the first mirror to the
second. At what angle with respect to the vertical does the light finally travel?

SOLUTION:

Procedure: Use the law of reflection. This means you must identify the normals and
the angles of incidence and reflection.

On Figure SG 21.1a are drawn the two normals as dotted lines and the two verticals as
light solid lines. The incident ray is indicated by the arrow AB, its reflection by BC and
the final reflected ray by CD. Angles α and β are used to help us determine the angle of
incidence. The first angle of incidence is γ and the first reflection is at an angle of δ with
respect to the normal. The light hits the second mirror at an angle λ with respect to the
normal of the second mirror and is reflected at an angle η with respect to the normal.
The angle we want, the angle between the ray and the vertical is ϕ. The angles $\psi, \mu,$ and
ρ are useful angles to help us with our calculations.

Argument to show that δ = 90° – tilt angle = 45°

Since the first mirror is at 45° with respect to the horizontal, we know that α
is also 45° and by similar sides we also know that β is also 45°.

In summary:
$$\text{tilt} = 45° \text{ implies } \alpha = 45°,$$
$$\text{which implies } \beta = 45°,$$
$$\text{which makes } \gamma = 45°.$$

The law of reflection

$$\text{angle of incidence } = \text{ angle of reflection}$$

gives

$$\delta = 45°.$$

Since $\gamma + \delta = 90°$, the line BC is horizontal. We have not worried about this when the initial drawing was made. Figure SG 21.1a is only a sketch to help us think about the problem. It is best not to draw figures at special angles (45°, 90°, etc.) unless you know that these are the correct angles. By avoiding these special angles you will avoid the temptations to make use of symmetries that do not exist.

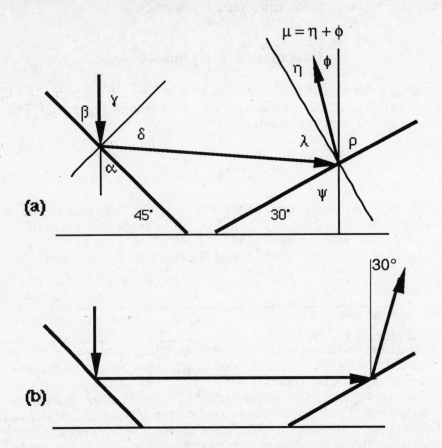

Figure SG 21.1: a). Two facing mirrors with light traveling vertically down-ward hitting the mirror on the left. A reflected light path is drawn which makes no assumptions about symmetry.
b). The correct light path.

Argument to show the angle between second normal and the horizontal, λ, is 90° − tilt.

We now need the normal to the second mirror. Since that mirror is at 30°, the angle ψ is 60° and the angle ρ must equal ψ and hence ρ is 60°. If ρ is 60°, then μ is 30° and the normal is tilted 60° with respect to the horizontal.

In general the normal always has a tilt λ equal to ψ. Since $\psi = 60°$

$$\lambda = 60°,$$

and, since $\lambda + \mu = 90°$,

$$\mu = 30°.$$

We have already determined that the incident ray is horizontal as it hits the second mirror, so the angle of incidence is 60°. This is a result of the particular geometry of this problem. In general, the incident angle is not the tilt of the normal.

The law of reflection gives
$$\lambda = \eta,$$
so
$$\eta = 60°.$$

The angle between the reflected ray and the vertical is the angle ϕ, and by inspection

$$\phi = \mu - \eta = -30°.$$

The minus sign indicates that ϕ lies on the opposite side of the vertical than we assumed in the drawing. Figure SG 21.1b shows our findings.

PROBLEM SG 21.2: Light is incident at 60° to the normal of one side of a glass prism whose apex angle is 30°. At what angle with respect to the normal does the light exit from the other side? Assume the index of refraction of the glass is 1.5.

SOLUTION:

Procedure: Use Snell's Law at each surface. The major problem is finding the final angle of incidence as the light passes from the glass out into the air.

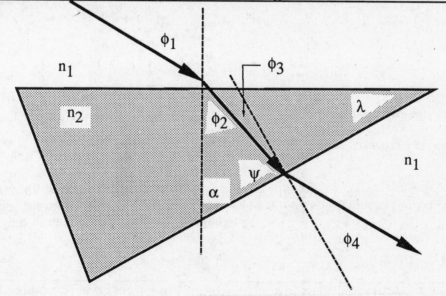

Figure SG 21.2: Light passing through a prism. The normals are shown dotted and the light ray as a thick arrow.

Figure SG 21.2 shows the geometry of the problem. The following notation is used:

Quantity	Symbol	Value
initial incident angle	ϕ_1	60°
apex angle of prism	λ	30°
first refracted angle	ϕ_2	—
second incident angle	ϕ_3	—
final refracted angle	ϕ_4	—
auxiliary angles	α, ψ	—
index of refraction of air	n_1	1.00
index of refraction of glass	n_2	1.50

Applying Snell's law at the first surface gives

$$n_1 \sin \phi_1 = n_2 \sin \phi_2$$

which can be rewritten as

$$\phi_2 = \sin^{-1}\left(\frac{n_1}{n_2} \sin \phi_1\right).$$

Likewise,

$$\phi_4 = \sin^{-1}\left(\frac{n_2}{n_1} \sin \phi_3\right).$$

The relation between ϕ_2 and ϕ_3 is found by noting that

$$\alpha = 90° - \lambda.$$

This must be true since α is the third angle of a right triangle. Then

$$\psi = 180° - \phi_2 - \alpha = 90° + \lambda - \phi_2$$

and

$$\phi_3 = 90° - \psi = \phi_2 - \lambda.$$

This gives

$$\phi_4 = \sin^{-1}\left\{\frac{n_2}{n_1} \sin\left[\sin^{-1}\left(\frac{n_1}{n_2} \sin \phi_1\right) - \lambda\right]\right\}.$$

Substituting the initial numbers gives $\phi_4 = 7°.9$.

The reflected ray emerges at an angle of 7°.9 with respect to the final normal.

PROBLEM SG 21.3: A thin lens with a focal length of 10 cm is located 25 cm to the left of another lens of focal length 5 cm. A 2 cm tall object is located 30 cm to the left of the first lens. What is the nature, location, and size of the final image? Draw the ray diagram.

SOLUTION:

Procedure: Apply the thin lens formula two times in succession. The image formed by the first lens serves as the object for the second lens.

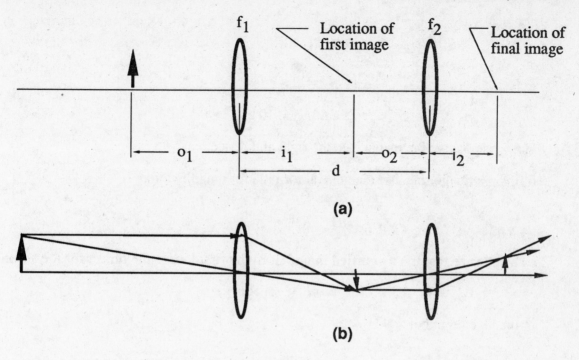

f_1 Location of first image f_2 Location of final image

(a)

(b)

Figure SG 21.3:   a). Two thin lenses.
                        b). The ray diagram.

The lenses and spacing are shown in Figure SG 21.3a. The following notation is used:

Quantity	Symbol	Value
left lens focal length	f_1	10 cm
right lens focal length	f_2	5.0 cm
separation of the lenses	d	25 cm
distance of object from 1st lens	o_1	30 cm
1st image distance	i_1	—
2nd object distance	o_2	—
final object distance	i_2	—
height of object	h	2 cm
height of final image	h_f	—
total magnification	m	—

We apply the thin lens formula to the first lens

$$\frac{1}{o_1} + \frac{1}{i_1} = \frac{1}{f_1}.$$

Our numbers give

$$i_1 = 15 \text{ cm}.$$

Since

$$i_1 + o_2 = d,$$

$$o_2 = 10 \text{ cm}.$$

Then using

$$\frac{1}{o_2} + \frac{1}{i_2} = \frac{1}{f_2}.$$

We find that

$$i_2 = 10 \text{ cm.}$$

Since i_2 is positive the image is to the right and hence real.

To find the image size, use the general formula for magnification

$$m = -\frac{i}{o}$$

and again note that the magnified image from the first lens is the object for the second lens, so

$$m = m_1 \times m_2.$$

For the given numbers

$$m = \left(-\frac{1}{2}\right) \times (-1) = \frac{1}{2}$$

Thus the image is real, erect ($m > 0$), 1 cm tall and is located 10 cm to the right of the second lens.

Figure SG 21.3b is a ray diagram. To keep the picture simple, only three rays are drawn. One undeviated ray for the tail of the arrow. Two rays for the tip according to the first and last rules in the text. Draw on this figure other lines according to the text rules for scale drawings.

PROBLEM SG 21.4: Repeat the previous problem with a lens separation of 16 cm.

SOLUTION:

Procedure: Repeat the same process as used in the preceding problem. Positive distances indicate real images, while negative distances will indicate virtual images.

Figure SG 21.3a and the table of symbols from the preceding problem may be used again. The only change is that

$$d = 16 \text{ cm.}$$

The value for i_1 is unchanged but

$$o_2 = 1.0 \text{ cm}$$

and from the thin lens formula

$$i_2 = -1.25 \text{ cm,}$$

so the image is 1.25 cm to the left ($i_2 < 0$) of the second lens. Thus the image is virtual.

The value of m_1 is unchanged and

$$m_2 = -\frac{i_2}{o_2} = -\frac{-1.25 \text{ cm}}{1.0 \text{ cm}} = 1.25$$

so

$$m = m_1 \times m_2 = -0.63.$$

The image is inverted, 1.25 cm tall, virtual, and located 1.25 cm to the left of the second lens. A ray diagram is shown in Figure SG 21.4.

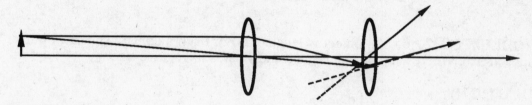

Figure SG 21.4:   Ray diagram for Problem SG 21.4.

To keep the picture simple, only three rays are drawn. One undeviated ray for the tail of the arrow. Two rays for the tip according to the first and last rules in the text.

PROBLEM SG 21.5: Repeat Problem SG 21.3 with the separation of the lenses 25 cm, as given in that problem, but with the second lens a diverging lens of focal length 5.0 cm.

SOLUTION:

> **Procedure:** Repeat the same process as used in the preceding problems. Positive distances indicate real images, while negative distances will indicate virtual images. A diverging lens has a negative focal length so $f_2 = -5.0$ cm.

The behavior of the light through the first lens is the same as in Problem SG 21.3 so the value of $o_2 = 10$ cm as before but now we must take $f_2 = -5.0$ cm. This gives

$$i_2 = -\frac{10}{3} \text{ cm},$$

so the image is virtual and about 3.3 cm to the left of the second lens. The magnification of the second lens is 1/3, so

$$m = m_1 \times m_2 = -\frac{1}{6} = -0.17.$$

The image is inverted, 0.33 cm tall, virtual, and located 3.3 cm to the left of the second lens. A ray diagram is shown in Figure 21.5.

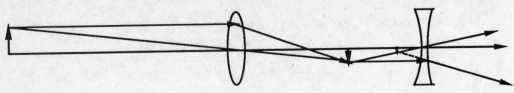

Figure SG 21.5:   Ray diagram for Problem SG 21.5

To keep the picture simple, only three rays are drawn. One undeviated ray for the tail of the arrow. Two rays for the tip according to the first and last rules in the text.

PROBLEM SG 21.6: An object is located 2.0 m from a convex spherical mirror with radius 0.80 m. Where is the image located and what is its relative size?

SOLUTION:

> **Procedure:** Use the thin lens formula with focal length equal to half the radius of the mirror. A convex mirror is diverging and has a negative radius.

Let R be the radius of the mirror, f its effective focal length, o the object distance, i the image distance, h the object height, and h_i the image height. Then

$$f = -\frac{R}{2} \quad \text{and} \quad \frac{1}{o} + \frac{1}{i} = -\frac{2}{R}.$$

The image size is given by

$$\frac{h_i}{h} = -\frac{i}{o}.$$

Entering numbers gives

$$i = -\frac{1}{3} \text{ m} \quad \text{and} \quad \frac{h_i}{h} = +\frac{1}{6}.$$

The image is erect and lies 0.33 m to the left of the mirror's surface. Remember that the sign convention for mirrors measures both image and object distance positive the lie to the same side of the mirror. This is shown in the ray diagram in Figure SG 21.6. Two rays are shown. Ray 1 travels parallel to the figure axis and is chosen so as to be reflected along a line through the focal point. Ray 2 passes through the center of the sphere defining the mirror and is reflected directly back.

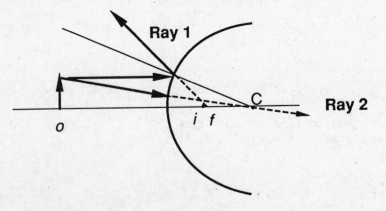

Figure SG 21.6:   Ray diagram for Problem SG 21.6

Chapter 22
Optical Instruments

Important Terms

Fill in the blanks with the appropriate word or words.

1. A(n) _____ lens is one whose focal length can be changed but which must be refocused at each setting.

2. A(n) _____ microscope consists of a lens close to the object under study and another lens near the observer's eye.

3. _____ is the eye's ability to focus on objects at various distances.

4. A(n) _____ glass consists of a single converging lens.

5. A(n) _____ lens is a variable focus lens which maintains its focus throughout its range.

6. The ratio of the focal length of a lens to the diameter of the lens is the _____ of the lens.

7. The lens of a microscope or telescope closest to the object is called the _____ lens.

8. The smallest distance from the unaided eye that an object can be held and still appear unblurred is the _____.

9. The _____ telescope has a mirror as its objective.

10. The lens of a microscope closest to the eye of the observer is called the _____.

11. If the human eye or other lens is distorted from its normal spherical shape, the lens is said to suffer from _____.

12. The ratio of the angular size of an object as seen through a lens to the angular size without the lens is called the _____ magnification.

13. A(n) _____ telescope has a diverging lens as its eyepiece.

14. The reciprocal of the focal length of a lens in meters is the strength of the lens in _____.

15. A(n) _____ telescope consists of two converging lenses (or two sets of lenses).

Write the definitions of the following:

16. astigmatism (22.1):

17. angular magnification (22.2):

18. objective lens (22.4):

19. refracting astronomical telescope (22.5):

20. zoom lens (22.6):

21. accommodation (22.1):

22. diopter (22.1):

23. f-number (22.3):

24. eyepiece (22.4):

25. Galilean telescope (22.5):

26. varifocal lens (22.6):

27. near point (22.1):

28. magnifying glass (22.2):

29. compound microscope (22.4):

30. ocular (22.4):

31. Newtonian telescope (22.5):

Answers to 1–15, Important Terms

1. varifocal; 2. compound; 3. accommodation; 4. magnifying; 5. zoom; 6. f-number;
7. objective; 8. near point; 9. Newtonian; 10. eyepiece (or ocular); 11. astigmatism;
12. angular; 13. Galilean; 14. diopters; 15. refracting (or inverting)

Sample Solutions

PROBLEM SG 22.1: A farsighted eye has a near point of 120 cm. A converging lens is used to provide clear vision of a book placed 25 cm in front of the eye. What is the power of the lens used in diopters?

SOLUTION:

> **Procedure:** The idea is to use the lens to produce a virtual image of the book at the near point. Use the thin lens equation. A correct choice of units will make the problem easier.

The idea of a corrective lens for farsightness is to place a converging lens near the eye. The power of the converging lens is chosen so as to produce an image at the near point of the eye when the object is held at a convenient distance. We let o be the distance the book is held, i be the location of the image (i is chosen to be the near point), f the focal length of the lens and p the power of the lens. This is shown in Figure SG 22.1.

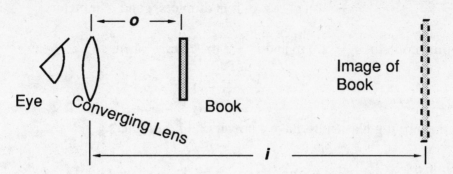

Figure SG 22.1: A converging lens used to allow reading at a convenient distance.

If we neglect the distance between the eye and the lens, we have

$$\frac{1}{o} + \frac{1}{i} = \frac{1}{f}.$$

If, in addition, all distances are measured in meters, we can also write:

$$\frac{1}{o} + \frac{1}{i} = p \qquad p \text{ in diopters, } i \text{ and } o \text{ in meters.}$$

For this problem $o = 0.25$ m and $i = -1.2$ m. The minus sign for i follows from the assumption that the object and the image are on the same side of the lens.

Then, entering the numbers, we get

$$p = +3.2 \text{ diopters} .$$

This corresponds to a focal length of 0.31 m = 31 cm. The lens will probably be within 2.5 cm of the eye, so our assumption that we could neglect the distance between the eye and the lens is reasonable.

PROBLEM SG 22.2: A magnifying glass held 2.0 m in front of a TV screen projects an image of the screen on a wall 3.0 m from the TV screen. What is the power of the lens?

SOLUTION:

Procedure: The lens produces a real image of the TV screen at the given distance. Use the thin lens equation. A correct choice of units will make the problem easier.

We let o be the distance between the lens and the TV screen i be the location of the image beyond the lens, f the focal length of the lens and p the power of the lens. This is shown in Figure SG 22.2.

The thin lens formula is

$$\frac{1}{o} + \frac{1}{i} = \frac{1}{f}.$$

If, in addition, all distances are measured in meters, we can also write:

$$\frac{1}{o} + \frac{1}{i} = p \qquad p \text{ in diopters, } i \text{ and } o \text{ in meters.}$$

For this problem, $o = 2.0$ m and $i = 3$ m $- 2$ m $= 1$ m, so

$$\frac{1}{2} + \frac{1}{1} = 1.5 = p.$$

The magnifying glass must have a power of $+1.5$ diopters.

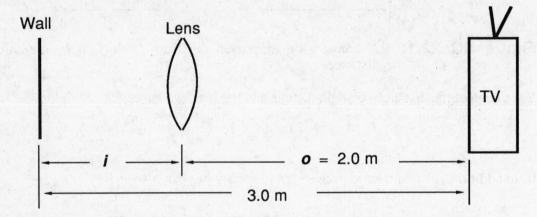

Figure SG 22.2:   Lens used to produce an image of a TV picture.

PROBLEM SG 22.3: A camera lens has a diameter of 2.0 cm and a focal length of 6.0 cm. Using a particular film, a proper exposure of a scene can be made with the camera set to an f-number of 5.6 and an exposure time of 1/60 s. Can the camera record the same scene with the same film at a speed of 1/500 s? What is the maximum shutter speed possible to properly expose this scene?

SOLUTION:

Procedure: Calculate the minimum f-ratio possible for this lens and compare that number with the f-number needed for the same exposure at 1/500 s.

Film responds to the total light hitting it per unit area during the exposure, so we require that

$$\frac{\text{exposure time}}{(\text{f-number})^2} = \text{constant}.$$

The f-number is adjusted by a diaphragm that reduces the open diameter of the lens, so the minimum f-number (which must occur with the diaphragm completely open) is given by

$$\text{f-number}_{min} = \frac{\text{focal length}}{\text{true diameter}} = 3.0.$$

To change the exposure time from 1/60 s to 1/500 s we must have

$$\frac{\text{exposure time}}{(\text{f-number})^2} = \frac{1/60 \text{ s}}{(5.6)^2} = \frac{1/500 \text{ s}}{(\text{f-number})^2} \, ,$$

which gives

$$\text{f-number} = 1.9.$$

Since 1.9 is less than the minimum possible f-ratio, it would be impossible to take the exposure at 1/500 s.

The minimum exposure time is given by

$$\frac{\text{exposure time}}{(\text{f-number})^2} = \frac{1/60 \text{ s}}{(5.6)^2} = \frac{t_{min}}{(3.0)^2} \, ,$$

which gives

$$t_{min} = \frac{1}{209} \text{ s.}$$

Since many cameras have an exposure setting of 1/200 s, that would probably be the best minimum time and hence the maximum shutter speed.

PROBLEM SG 22.4: The maximum useful magnification of an ordinary visible light microscope is about 2000x. A microscope comes with a set of eyepieces the most powerful of which is marked 25x. Assuming that the manufacturer gave you the maximum useful eyepiece, what is the focal length of the objective used to get the maximum magnification.

SOLUTION:

Procedure: Assume the standard spacing of 16 cm for the microscope body and use the relationship between the magnification of a microscope and the eyepiece used.

Let M be the magnification of the microscope, M_e be the magnification of the eyepiece, L the body length of the microscope, and f_o the focal length of the objective. We can write the relation among these as:

$$M = -\frac{L}{f_o} M_e \, .$$

Take $M = -2000$x, $M_e = 25$x, and $L = 16$ cm to find

$$f_o = 0.20 \text{ cm.}$$

The objective of the microscope has a focal length of 2.0 mm.

PROBLEM SG 22.5: You are watching a basketball game with a pair of 7×50 binoculars. If a 2.0 m tall basketball player is 50 m away from you, what is his apparent angular size in radians as seen through the glasses?

SOLUTION:

Procedure: Compute the true angular size of the player and use the definition of magnification.

The magnification of a pair of binoculars is defined as

$$M = \frac{\text{angular size with aid}}{\text{angular size without aid}} \cdot$$

Since we are given $M = 7$, we need the true angular size in order to calculate the apparent angular size.

For an object whose size is small compared with its distance,

$$\text{angular size without aid} = \frac{\text{true size}}{\text{distance}} = \frac{2.0 \text{ m}}{50 \text{ m}} = 0.040 \text{ radians} .$$

The angular size as seen through the binoculars is just 7 times this, so the player seem to intercept an angular size of 0.28 radians.

Chapter 23
Wave Optics

Important Terms

Fill in the blanks with the appropriate word or words.

1. A(n) _____ converts light to electrical energy and displays the intensity versus wavelength on a chart.

2. Sources with identical frequency and fixed relative phase are said to be _____.

3. The _____ surface (or _____ front) is defined as the set of points of constant phase.

4. A(n) _____ is a device to view the spectrum of a light source with the eye.

5. An array of a large number of equally spaced slits is called a(n) _____.

6. _____ shows itself in the wavelength dependence of the index of refraction.

7. The plane of _____ of light is the plane of the electric field.

8. _____ Law relates the intensity of light passing through two polarizers to the angle between their planes of polarization.

9. When light is _____ by dust, the light is absorbed by the dust grains and then re-emitted in an arbitrary direction.

10. _____ are inexpensive plastic polarizers.

11. The angle of maximum polarization of reflected light is given by _____ Law.

12. A(n) _____ makes a photographic record of a spectrum.

13. The rotation of the plane of polarization of incident light when the light passes through a substance is an example of _____.

14. The ability to construct a new wave front by drawing arcs centered on the previous wave front is called _____ principle.

15. The _____ of an optical instrument is a measure of that instrument's ability to produce a distinct image of two objects which appear very close together.

16. Scattering of light by particles smaller than the wavelength is called _____ scattering.

17. Rayleigh's _____ is to choose the minimum (best) resolution as that angle at which the first minimum of the diffraction pattern of one source falls on the maximum of the other source.

18. The _____ of light is shown by light spreading into regions which should be shadowed from the light.

19. A(n) _____ coating on a lens is made of a material with an index of refraction between that of air and that of glass.

Write the definitions of the following:

20. Malus's Law (23.10):

21. wave front (or wave surface) (23.1):

22. diffraction (23.5):

23. interference (23.3):

24. antireflection coating (20.4):

25. dispersion (23.8):

26. optical activity (23.11):

27. Rayleigh scattering (23.10):

28. coherent (23.3):

29. polarization (23.9):

30. resolving power (20.9):

31. spectrograph (23.9):

32. spectroscope (23.9):

33. scattering (23.11):

34. Rayleigh's criterion (20.7):

35. Brewster's Law (23.9):

36. Huygen's Principle (23.1):

37. diffraction grating (23.6):

38. spectrophotometers (23.8):

Answers to 1-19, Important Terms

1. spectrophotometer; 2. coherent; 3. wave; 4. spectroscope; 5. diffraction grating; 6. dispersion; 7. polarization; 8. Malus's; 9. scattered; 10. polaroids; 11. Brewster's; 12. spectrograph; 13. optical activity; 14. Huygen's; 15. resolving power; 16. Rayleigh; 17. criterion; 18. diffraction; 19. antireflection

General Comments

Light is modeled as a transverse wave. Ripples on the surface of water provide an everyday experience of a transverse wave. Our model of light differs from that of water ripples in that there are two transverse parts to a light wave — the electric field and the magnetic field. These are at right angles to each other and each obeys the traveling wave equation of Chapter 14. If the plane of the electric field (and hence the magnetic field) is fixed in space we have a plane polarized wave. In general the plane is not fixed and obeys no simple time behavior; we call such light unpolarized. There are other types of polarization such as elliptical and circular which are not discussed in the text.

We call the wave picture of light a model for two reasons. First, we have not investigated the fundamental properties of the electric and magnetic fields that produce the wave and that govern the interaction of a light wave with matter. Second, as we will see in Chapter 24, there are times when light acts as a particle. The properties of light discussed in this chapter can be most easily understood by considering light as a wave. There are other properties which can only be easily understood if light is a particle. This dual property will be more fully explained in Chapters 24 and 25.

As indicated in the text, the wave property of diffraction follows from Huygen's Principle. In this sense we can consider diffraction as self interference of a single wave front. For a two slit interference pattern we examine the relative phase of light from the two different slits. For a diffraction grating we examine the relative phases from the set of slits. For a diffraction pattern we examine the relative phases of the light that passed through various parts of a single slit.

Sample Solutions

QUESTION SG 23.1: A water proof flashlight is dropped into a pond. It lands in the muck at the bottom, sticking up at an angle of 30° from the vertical, and at a depth of 1.5 m from the surface, with the light on. The flashlight is viewed from outside the pond at a height such that it appears to be seen end on. From this position it appears that the flashlight is 4.90 m horizontally from the edge of the pond. How far away from the edge of the pond is the flashlight?

SOLUTION:

> **Procedure:** Apply the law of refraction at the surface of the water. This will enable you to calculate the apparent position of the flashlight as seen by the observer.

We sketch the problem in Figure SG 23.1. Since the flashlight appears to be seen end on, a light ray leaving the flashlight normally is traveling at an angle of 30° to the vertical, is refracted at the air-water interface and reaches the observers eyes as shown. The following notation is used:

Quantity	Symbol	Value
angle of incidence	θ	30°
index of refraction of water	n_{water}	1.33
index of refraction of air	n_{air}	1.00
angle of refraction	ϕ	unknown
depth of flashlight	h	1.5 m
horizontal distance to flashlight	d	unknown
horizontal distance from source to where the observed light leaves the water	z	unknown
apparent displacement of light	x	unknown
apparent horizontal distance to flashlight	D	4.90 m

Of the unknown items listed, we are asked to find the true position of the flashlight, d. The other quantities will be needed as part of the intermediate steps. The approach will be to use Snell's law at the interface to find the angle of refraction, ϕ, which will allow us to calculate the distance $x+z$. The distance z can also be determined from the given information so we can clearly find d.

As is shown in Figure SG 23.1a, the light that is seen by the observer travels from the source to the point marked A, is refracted and then travels to the observer's eyes. To the observer it appears that the light traveled in a straight line as shown by the dotted line. Knowing that the pond is 1.5 m deep we can see from Figure SG 23.1b that

$$\tan \theta = \frac{z}{h} = \tan 30°,$$

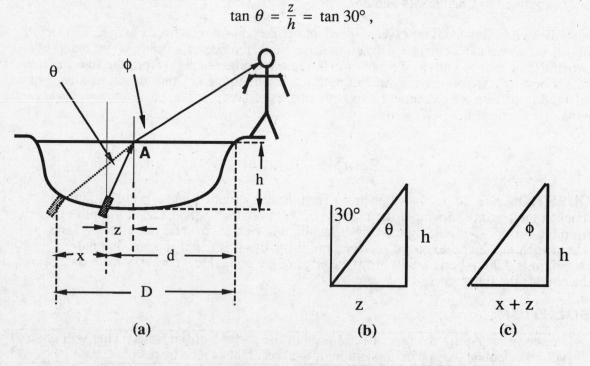

(a) **(b)** **(c)**

Figure SG 23.1:   a). Geometry and light path for flashlight problem.
b). Geometry to determine z.
c). Geometry to determine $x + z$.

and from Figure SG 23.1c

$$\tan \phi = \frac{x + z}{h}.$$

Hence

$$x = h \left\{ \tan \phi - \tan 30° \right\},$$

and the answer is

$$d = D - x = D - h \left\{ \tan \phi - \tan 30° \right\}.$$

We only need to find ϕ, which can be done by using Snell's Law at point A.

$$n_{water} \sin \theta = n_{air} \sin \phi$$

or

$$1.33 \times \sin 30° = 0.665 = 1.00 \times \sin \phi,$$

hence

$$\phi = \sin^{-1}(0.665) = 41°.9.$$

Since we need $\tan \phi$, it is best to do the calculation as one step on the calculator and get

$$\tan \phi = \tan \left\{ \sin^{-1}(0.665) \right\} = 0.89$$

hence

$$d = 4.90 \text{ m} - 1.5 \text{ m} \times (0.89 - 0.58) = 4.4 \text{ m}$$

The flashlight has fallen 4.4 m from the edge of the pond.

PROBLEM SG 23.2: A pair of narrow slits are separated 0.101 mm from each other. Light of wavelength 567 nm passes through the slits and falls on a screen 2.5 m away from the slits and parallel to the plane of the slits. What is the linear separation of the central maximum and either of the second minima?

SOLUTION:

> **Procedure:** The formulae for double slit maxima and minima give us the angle to the maxima and minima as seen from the slit. This angle and the distance to the screen allow us to calculate the linear separation of the central maximum and a second minimum.

We draw the problem in Figure SG 23.2 and let d be the slit separation, D the slit to screen distance, x the linear distance from the central maximum to the second minima and λ be the wavelength.

From the figure we see that x and D are related to the shown angle ϕ by

$$\tan \phi = \frac{x}{D}.$$

We have only shown a positive x. There is another symmetric position for negative ϕ and hence negative x with the same magnitude of ϕ.

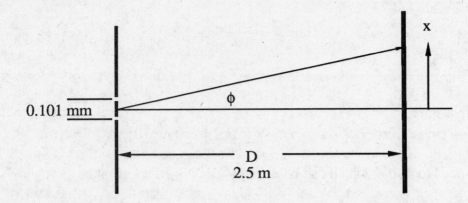

Figure SG 23.2: Light from two slits falling on a screen 2.5 m away.

If we let θ represent a general value of ϕ, the maxima for this double slit are given by (text Equation 23.3a)

$$n\lambda = d \sin \theta \qquad \text{for } n = 0, 1, 2, \ldots$$

and the minima by (text Equation 23.3b)

$$\left(m + \frac{1}{2}\right)\lambda = d \sin \theta \qquad \text{for } m = 0, 1, 2, \ldots$$

The central maximum is $n = 0$ or $\theta_{\text{central max}} = 0$. The second minima are given by $m = 1$, so the angle to a second minimum, ϕ, is given by

$$\left(1 + \frac{1}{2}\right)\lambda = d \sin \phi.$$

Since $\theta_{\text{central max}} = 0$, the angle denoted by ϕ is also the angle shown in Figure SG 20.2. Thus

$$x = D \times \tan \phi = D \times \tan\left\{\sin^{-1}\left(\frac{3\lambda}{2d}\right)\right\}$$

$$x = 2.5 \text{ m} \times \tan\left\{\sin^{-1}\left(\frac{3 \times 567 \times 10^{-9} \text{ m}}{2 \times 1.01 \times 10^{-4} \text{ m}}\right)\right\}$$

$$= 0.021 \text{ m} = 21 \text{ mm}.$$

The second minima are 21 mm from the central maximum.

PROBLEM SG 23.3: A horizontal glass sheet is covered by thin water droplets containing a small amount of wetting agent. The wetting agent allows the water droplets to have a large horizontal area as shown in Figure SG 23.3. How many molecules thick must a water droplet be for normally incident light of wavelength 500 nm to have a maximal reflection? Assume that the index of refraction of water is unchanged by the wetting agent (n = 1.33) and that a water molecule occupies a volume with a diameter of about 0.5 nm.

SOLUTION:

> **Procedure:** Some light is reflected off the water and some passes through the water and is reflected off the glass. The optical path difference must be that given for constructive interference.

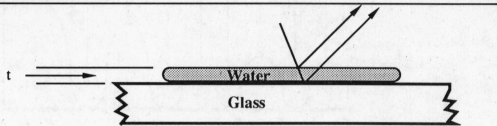

Figure SG 23.3:   Thin layer of water on a horizontal glass plate.

We assume that light is reflected off both the water and the glass as shown in Figure SG 23.3. The incident ray is shown at an angle rather than the normal to show the paths taken by the light. Since the index of refraction of air is less than that of water and since the index of refraction of water is less than that for glass, constructive interference will occur for

$$m\lambda = 2nt,$$

where m is an integer 0, 1, 2, ..., and λ is the wavelength of light, n is the index of refraction of water and t is the water thickness (see text Equation 23.5a). We assume that $m \neq 0$, since m = 0 would correspond to no water,. So we take $m = 1$. Then the thickness of the water is

$$t = \frac{\lambda}{2n} = 188 \text{ nm}.$$

Since each water molecule occupies about 0.5 nm, the water layer is about

$$\frac{t}{0.5 \text{ nm}} = 4 \times 10^2 \text{ molecules thick}.$$

PROBLEM SG 23.4: A pair of narrow slits are used to generate an interference pattern. The intensities of the maxima decrease through the 3rd order and then increase for the 4th order. This suggests that the 1st diffraction minima are near the 2nd order interference minima. Assuming that the 1st diffraction minima lie on the 2nd order interference minima, what is the ratio of the slit width to the slit separation?

SOLUTION:

> **Procedure:** Apply the relations giving maxima and minima for two slit interference patterns. The main difficulty with many problems such as these is to determine which minima and maxima are being discussed.

Figure SG 23.4 shows the left hand half of the pattern with the numbering of the maxima and minima.

We let λ be the wavelength of light used, d the slit separation, b the slit width, and θ indicate an angle.

Interference maxima

Interference minima

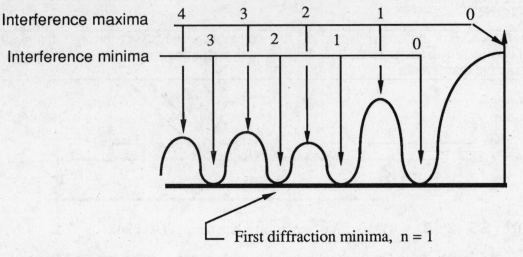

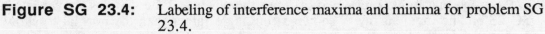

First diffraction minima, n = 1

Figure SG 23.4: Labeling of interference maxima and minima for problem SG 23.4.

For an interference minimum

$$\left(m + \frac{1}{2}\right)\lambda = d \sin \theta_m \qquad m = 0, 1, 2, \dots$$

while for a diffraction minimum

$$n\lambda = b \sin \theta_n \qquad n = 1, 2, 3, \dots$$

For this problem $n = 1$ (1st diffraction minima) and $m = 2$ (3rd interference minima), so

$$\frac{\lambda}{b} = \sin \theta_n = \sin \theta_m = \frac{5\lambda}{2d} .$$

This gives

$$\frac{b}{d} = \frac{2}{5} = 0.4.$$

Each slit is 0.4 times as wide as the slit separation.

PROBLEM SG 23.5: Light passing through two aligned polarizers has an intensity of 3.7×10^{-5} W/m². Through what angle must one of the polarizers be turned in order to reduce the light intensity to 7.0×10^{-6} W/m²?

SOLUTION:

Procedure: This is a straightforward application of Malus's Law.

If we let θ be the angle through which we turn one polarizer as shown in Figure SG 23.5, I_m the initial maximal intensity and I the intensity after turning through θ, we have

$$I = I_m \cos^2 \theta .$$

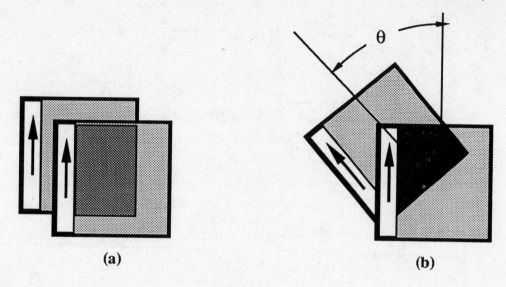

(a) **(b)**

Figure SG 23.5:   a). Two aligned polarizers.

b). One polarizer now rotated through an angle θ.

Solving Malus's Law for θ gives

$$\theta = \cos^{-1}\left(\sqrt{\frac{I}{I_m}}\right) = \cos^{-1}\left(\sqrt{\frac{7.0 \times 10^{-6}}{3.7 \times 10^{-5}}}\right) = 64°.$$

One of the polarizers must be turned through an angle of 64°.

PROBLEM SG 23.6: Two identical circular apertures have been cut in a copper foil. The center to center separation of the apertures is twice the diameter of either aperture. The apertures are passing light of 500 nm wavelength. and are viewed from a distance of 100 m. If they are just barely resolvable as two apertures, what is the diameter of either aperture?

SOLUTION:

> **Procedure:** We can express the true angular separation of the apertures in terms of the distance and the separation of the holes. Using Rayleigh's criteria, we can also relate this angular separation to the ratio of the wavelength to the diameter

If we let D be the diameter of either aperture and L be the distance to the plate as shown in Figure SG 23.6, then the angular separation of the apertures, θ is given by

$$\theta = \frac{\text{separation}}{L} = \frac{2 \times D}{L}.$$

Use Rayleigh's criterion and equate θ to θ_m in text Equation 23.10 to get

$$\frac{2 \times D}{L} = \theta = \theta_m = 1.22 \frac{\lambda}{D}.$$

This gives

$$D = \sqrt{0.61\, \lambda\, L} = 5.5 \times 10^{-3}\ m = 5.5\ mm.$$

The holes have a diameter of 5.5 mm.

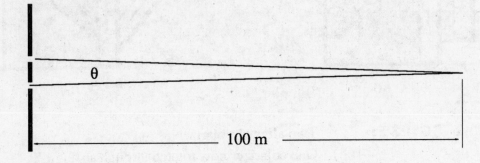

Figure SG 23.6: Two holes viewed from a distance of 100 m.

Chapter 24
Relativity

Important Terms

Fill in the blanks with the appropriate word or words.

1. The _____ theory of relativity treats accelerated reference frames and gravity.

2. The idea that being held at rest in a gravitational field is the same as uniform acceleration is called the _____.

3. A non-accelerating frame of reference is called a(n) _____ frame.

4. The _____ theory of relativity resolved the conflicts between electromagnetic theory and mechanics for problems that do not involve gravity.

5. The time that a clock measures as seen from the clock's rest frame is called the clock's _____ time.

6. The measured length of a moving rod is different than its rest length.This effect is known as a(n) _____ contraction.

7. The equations relating the coordinates in one reference frame to those of another reference frame moving with constant velocity relative to the first frame are called the _____ equations.

8. The _____ mass of a body is the body's mass as measured in a frame at rest with respect to the body.

9. The relation that exists between the mass of a body and the energy contained in that body is called the _____ equation.

10. The _____ length of an object is the object's length as measured in its rest frame.

11. Newton's Laws transform according to the rules of _____ relativity.

12. Two events are _____ to an observer, if in the observer's reference frame the events occur at the same time.

13. The time intervals that an observer measured with a clock moving relative to that observer are smaller than the intervals that would be measured if the clock were at rest relative to the observer. This effect is known as time _____.

14. _____ physics contains the ideas known at the end of the 19th century.

15. _____ is the concept of two event occurring at the same time.

Write the definitions for the following:

16. principle of equivalence (24.11):

17. Galilean relativity (24.1):

18. inertial (reference) frame (24.1):

19. simultaneous (24.4):

20. length contraction (24.6):

21. proper length (24.6):

22. time dilation (24.5):

23. proper time (24.5):

24. mass-energy relation (24.7):

25. general theory of relativity (24.12):

26. special theory of relativity (24.2):

27. (Lorentz) transformation equations (24.4):

28. rest mass (24.8):

Answers to 1-15, Important Terms

1. general; 2. principle of equivalence; 3. inertial; 4. special; 5. proper; 6. length (or Lorentz); 7. Lorentz transformation; 8. rest; 9. (Einstein) mass–energy; 10. proper; 11. Galilean; 12. simultaneous; 13. dilation; 14. Classical; 15. Simultaneity

General Comments

Both special and general relativity provide new models of physical reality. First let us examine the differences between Newton theory and the special theory relativity:

NEWTONIAN	SPECIAL RELATIVITY
space and time are separate entities	space and time are aspects of a single entity
space is the same for all observers time is the same for all observers	the split between space and time is made differently by different observers

If we are dealing only with velocities which are small compared to the speed of light, the differences between classical (Newtonian) physics and special relativity become too small to measure and the predictions of either are essentially the same. This reduction is a physical requirement for any new model of the world — it must include the old model in some limit.

Special relativity leads to many surprising predictions but it does not lead to paradoxes. A **paradox** is when two mutually contradictory conclusions are reached. A surprising result is not necessarily a paradox. An example of a paradox would be that a door is observed to be both open and shut at the same time. Many problems posed in special relativity may look paradoxical until the problems are carefully examined. Several of the sample problems will examine apparent paradoxes.

The time dilation and the length contraction of special relativity are not paradoxical since both ideas involve measurements made at two separate locations at the same time and hence involve simultaneity. These effects should not be called "apparent effects only" since the observed length has all the properties we expect length to have. The contraction and dilation are real to the observer. This is one of the essential "new" properties of the relativity model. Properties extending in space and time depend upon the observer. This dependence upon the observer is even more important when we consider quantum mechanics in later chapters.

Surprising results should do disconcert us too much as long as our ideas of causality are preserved. Let us define an event as a single position and time in one frame. In any other moving frame the event has a single position and time also. The number will, of course, depend upon the frame. A photoflash bulb firing is an example of an event. The flash occurs at a single location and time.

If event **A** "causes" event **B**, then the time order of **A** and **B** should be the same to all observers. For example the photoflash firing causes my eyes to hurt. No moving observer should claim that my eyes hurt before the photoflash bulb was fired.

Special relativity preserves this idea of causality; however, if two events are not causality related then it is possible that the time order will be reversed to two different observers. There is even an observer for which the two events are simultaneous. We will see how this property relates to the measurement of lengths in the sample problems.

In general relativity the space-time of special relativity becomes a dynamical entity of the model. That is, space and time itself are modified by the presence of matter. Again, general relativity must reduce to special relativity when the gravitational fields become weak. A measure of the strength of a gravitational field for this model is the quantity

$$\frac{2Gm}{Rc^2},$$

where G is the Newtonian gravitational constant, m is the mass of the object, R is the size of the object and c is the speed of light. When this quantity is much less than one, then special relativity holds.

In summary:

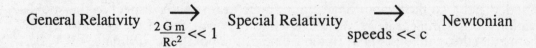

General Relativity $\xrightarrow[\frac{2\,G\,m}{Rc^2} \ll 1]{}$ Special Relativity $\xrightarrow[\text{speeds} \ll c]{}$ Newtonian

Sample Solutions

PROBLEM SG 24.1: A meter stick ($L_0 = 1.00$ m) is moving at 0.90 c relative to us. Verify that its contracted length is less than 0.5 m. This being the case it is obvious that if we had an open box of length 0.9 m in the path of the moving stick, as shown in Figure SG 24.1a, we could close the open side and have the stick entirely inside the box as shown in Figure SG 24.1b for the short time before the stick hits the back side of the box. But from the frame of the stick it is the box which is Lorentz contracted, so clearly the stick is never completely inside the box. What is the resolution of this apparent paradox? Hint: examine the problem in the frame of the meter stick.

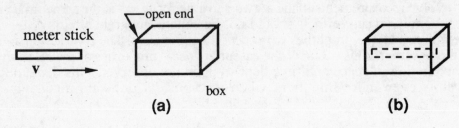

meter stick

open end

box

(a)

(b)

Figure SG 24.1: a). Meter stick approaching a 0.9 m box at a speed of 0.99 c.
b). Meter stick inside box.

SOLUTION:

Procedure: Use the Lorentz length contraction to calculate the length of the meter stick as it appears in the rest frame of the box. Then use the transformation equations to express what happens in the frame of the meter stick to resolve the apparent paradox.

The length of the moving meter stick can be calculated by the Lorentz contraction formula Equation 24.3. The measured length, L, will be

$$L = L_0 \sqrt{1 - \frac{v^2}{c^2}} = 1.00 \text{ m} \sqrt{1 - 0.90^2}$$

$$= 1.00 \text{ m} \times 0.43 = 43 \text{ cm},$$

which is clearly less than 0.9 m.

To resolve the apparent paradox we need to carefully examine the events involved. There are three events which we will call **A**, **B** and **C**.

Event **A** is when the front end of the meter stick just enters the box.
Event **B** is when the back end of the meter stick enters the box and
event **C** is when the front end of the stick hits the back end of the box.

These three events are shown in Figure SG 24.2 as measured by the observer at rest with respect to the box.

Let x and t be the space and time coordinates measured in a frame at rest with respect to the box and x' and t' be those coordinates in a frame at rest with respect to the stick. Our equations look simplest if we choose $x = 0$ at the open end of the box and $t = 0$ when the meter stick enters the box. In the frame of the box the coordinates of **A**, **B**, and **C** are given by

$$x_A = 0, t_A = 0 \text{ for } A,$$

$$x_B = 0, t_B = \frac{L}{v} = 1.6 \times 10^{-9} \text{ s for } B,$$

and

$$x_C = 0.9 \text{ m}, t_C = \frac{0.9 \text{ m}}{v} = 3.3 \times 10^{-9} \text{ s for } C,$$

where L is the measured length of the meter stick in the frame of the box and v is the velocity of the stick.

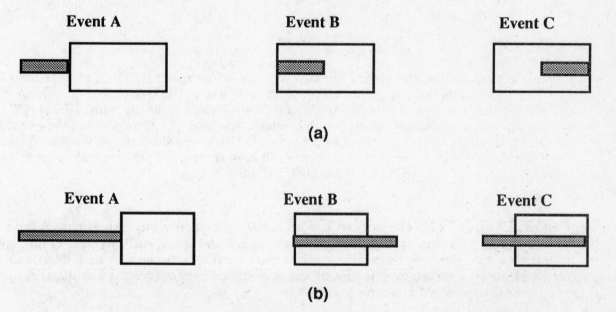

(a)

(b)

Figure SG 24.2: a). The three events as measured in the frame of the box.
b). The three events as measured in the frame of the meter stick.

We can calculate these events using the Lorentz transformation Equations (24.2). For event A:

$$x'_A = \frac{x_A - v t_A}{\sqrt{1 - \frac{v^2}{c^2}}} = 0 \text{ m};$$

$$t'_A = \frac{t_A - \frac{v x_A}{c^2}}{\sqrt{1 - \frac{v^2}{c^2}}} = 0 \text{ s}.$$

For event **B**:

$$x'_B = \frac{x_B - v t_B}{\sqrt{1 - \frac{v^2}{c^2}}} = -1.0 \text{ m};$$

$$t'_B = \frac{t_B - \frac{vx_B}{c^2}}{\sqrt{1 - \frac{v^2}{c^2}}} = 3.7 \times 10^{-9} \text{ s.}$$

For event **C**:

$$x'_C = \frac{x_C - vt_C}{\sqrt{1 - \frac{v^2}{c^2}}} = 0.00 \text{ m;}$$

$$t'_C = \frac{t_C - \frac{vx_C}{c^2}}{\sqrt{1 - \frac{v^2}{c^2}}} = 1.4 \times 10^{-9} \text{ s.}$$

In the frame of the meter stick, event **C** occurs before event **B**. Thus, as measured in the rest frame of the stick, the stick hits the back side of the box before the rear of the stick enters the box. There is no paradox, only the unusual, in terms of every day experience, disagreement as to which event occurs first. The time order of the event **B** and **C** can be reversed because they are separated by a time shorter than the time it takes for light to travel between the two events. In both frames the physics makes sense and either observer can predict what the other will measure.

PROBLEM SG 24.2: The star α-Centauri is 4.3 light years away. An astronaut travels to this star at 0.85 c and returns at 0.80 c where c is the speed of light. Neglecting any problems with turning around and neglecting the time it takes to turn around, describe the elapsed time as observed from the Earth and as observed by the traveling astronaut. A light year is the distance light to travels in one year.

SOLUTION:

Procedure: Calculate the elapsed time as measured by an observer on the Earth; then, use the time dilation relation to find the traveler's elapsed time.

We will approach this problem be using the relation between velocity and distance to calculate the elapsed time as measured from the Earth. We can then use the equation for time dilation (text Equation 24.4) to calculate the elapsed time in the astronaut's frame. When we use text Equation 24.4, remember that Δt_0 is the elapsed time in the rest frame of the clock and Δt is the elapsed time as measured in a frame with relative velocity v.

The total elapsed time as seen from the Earth is given by

$$\text{elapsed time} = \frac{\text{distance to α-Cent}}{\text{speed out}} + \frac{\text{distance to α-Cent}}{\text{speed back}}$$

$$= \frac{4.3 \text{ ly}}{0.85 \text{ c}} + \frac{4.3 \text{ ly}}{0.80 \text{ c}}$$

$$= (5.1 + 5.4) \text{ years} = 10.5 \text{ years.}$$

We can calculate the elapsed time (Δt_0 in text Equation 24.4) as seen by the astronaut:

$$\text{elapsed time} = (\text{time out in Earth frame}) \times \sqrt{1 - \frac{v_{out}^2}{c^2}}$$

$$+ (\text{time back in Earth frame}) \times \sqrt{1 - \frac{v_{back}^2}{c^2}}$$

$$= (2.7 + 3.2) \text{ years} = 5.9 \text{ years}.$$

On this trip 5.9 years will elapse to the astronaut while 10.5 years pass on Earth.

PROBLEM SG 24.3: A spaceship is approaching α-Centauri at 0.95 c. The maximum of energy of the light from this star, in the frame of the star, is centered on a wavelength of about 5.5×10^{-7} m which is in the center of the visible spectrum. What is the wavelength of this light as seen from the spaceship. What does this suggest would be one of the problems of space travel, if we could reach speeds near the speed of light?

SOLUTION:

Procedure: Use the Doppler formula to calculate the observed wavelength. Then, consider what the observed wavelength implies.

The problem as stated is illustrated in Figure SG 24.3a. As seen from the spaceship, the star is approaching at a speed v = 0.95 c as shown in Figure SG 24.3b.

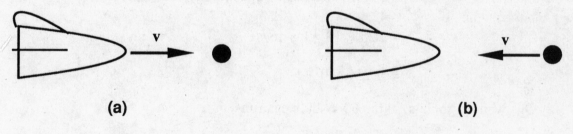

      (a)   **(b)**

Figure SG 24.3:   a). Spaceship approaching star at speed v.
                                 b). Picture as measured in frame fixed on the spaceship.

We let f be the observed frequency, f_0 be the emitted frequency and v the relative velocity. The observed wavelength is $\lambda = c/f$ and the emitted wavelength $\lambda_0 = c/f_0$. Then, since the velocity is one of approach, we can use text Equation (24.12):

$$f = \frac{c}{\lambda} = \frac{f_0 \sqrt{1 - \frac{v^2}{c^2}}}{1 - \frac{v}{c}} = \frac{\frac{c}{\lambda_0} \sqrt{1 - \frac{v^2}{c^2}}}{1 - \frac{v}{c}}.$$

Solving this for the wavelength, λ, observed gives:

$$\lambda = \frac{\lambda_0 \left(1 - \frac{v}{c}\right)}{\sqrt{1 - \frac{v^2}{c^2}}}$$

$$= \frac{5.5 \times 10^{-7} \text{ m} \times 0.05}{0.31} = 8.9 \times 10^{-8} \text{ m}.$$

This wavelength is much shorter than visible light (visible light lies in the range 4.4 – 7.0×10^{-7} m). The value of λ is hard ultraviolet radiation. To protect from this harmful radiation, the spaceship will need shielding. Shielding means extra mass which will make it harder to get the spaceship to a high velocity.

PROBLEM SG 24.4: If 2.50 mg of matter were completely converted to energy, how long could this energy run a 60 W light bulb?

SOLUTION:

> **Procedure:** Use the Einstein mass-energy relation to determine the energy available and the definition of the watt to calculate the time.

The total energy contained in an amount of mass m is given by the Einstein mass energy relation $E = mc^2$, where c is the speed of light. This energy will be expressed in joules. For 2.50 mg we have

$$E = m c^2 = 2.5 \times 10^{-6} \text{ kg} \times (3.00 \times 10^8 \text{ m/s})^2$$

$$= 2.25 \times 10^{11} \text{ kg-m}^2/\text{s}^2$$

$$= 2.25 \times 10^{11} \text{ J}$$

One watt = 1 joule-s, so the 60 W bulb can run

$$\frac{2.25 \times 10^{11} \text{ J}}{60 \text{ W}} = 3.8 \times 10^9 \text{ seconds} = 1.0 \times 10^6 \text{ hr} \approx 1.2 \times 10^2 \text{ years}.$$

The 60 W bulb can be run for about 120 years of the energy in 2.5 mg of matter.

PROBLEM SG 24.5: If we use the relativistic mass given by text Equation (24.9) and measure forces and accelerations in the frame of the observer (not in the rest frame of the particle), the equation $F = ma$ still applies. An electron moving through an electric field of 1.00×10^4 V/m (directed parallel to the electron's velocity) is observed to undergo an acceleration of 1.05×10^{15} m/s^2. What is the velocity of this electron? Hint: the electronic charge should be considered to be the same in all frames.

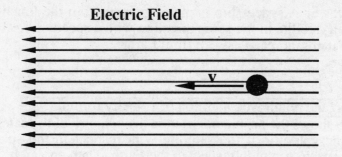

Figure SG 24.4:   Electron moving parallel to electric field.

SOLUTION:

> **Procedure:** Use $F = ma$ with F the electric force to calculate the effective mass. Use the effective mass and the rest mass to get the velocity.

The observed mass of the electron is given by

$$m = \frac{F}{a}$$

where F will be the electric force and a is the measured acceleration. The electric force is given by

$$F = Ee$$

where is the charge on the electron and E is the electric field. Thus, the measured mass, m, is given by

$$m = \frac{eE}{a} = \frac{1.60 \times 10^{-19} \text{ C} \times 1.00 \times 10^4 \text{ V/m}}{1.05 \times 10^{15} \text{ m/s}^2}$$

$$= 1.52 \times 10^{-30} \text{ kg.}$$

This measured mass, m, is related to the rest mass, m_0, by text Equation (24.8):

$$m = \frac{m_0}{\sqrt{1 - \frac{v^2}{c^2}}}.$$

Solving this equation for v gives us:

$$v = c \times \sqrt{1 - \frac{m_0^2}{m^2}},$$

so

$$v = 3.00 \times 10^8 \text{ m/s} \times \sqrt{1 - \left(\frac{9.11 \times 10^{-31}}{1.52 \times 10^{-30}}\right)^2}$$

$$= 2.4 \times 10^8 \text{ m/s} = 0.80 \text{ c.}$$

The electron is moving at 0.80 c.

PROBLEM SG 24.6: A space ship is moving away from the Earth at a speed of 0.50 c when it launches a satellite in the same direction with a speed of 0.60 c, as seen from the ship. What is the satellite's speed as seen from Earth?

SOLUTION:

Procedure: Use the relativistic addition of velocity formula.

Figure SG 24.5 shows the problem. Let u be the velocity of the satellite with respect to the Earth, u' the velocity of the satellite with respect to the space ship. and v the velocity of the space ship with respect to the Earth. Then a straightforward application of text Equation 21.1 gives:

$$u = \frac{u' + v}{1 + \frac{u'\,v}{c^2}}$$

$$= \frac{0.5c + 0.6c}{1 + \frac{(0.5c)(0.6c)}{c^2}} = \frac{1.1 \times c}{1 + 0.30} = 0.85\ c.$$

The satellite moves away from the Earth with a speed of 0.85 c as measured by an observer on Earth — less than the result predicted by the classical addition of relative velocities.

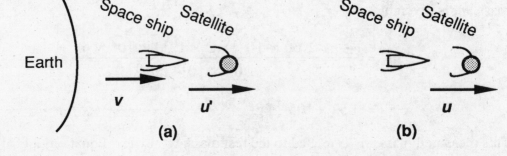

Figure SG 24.5:   a). Space ship launching a satellite. as seen from the Earth.
b). Launch as seen from the space ship.

PROBLEM SG 24.7: An electron is accelerated from rest to a velocity $v = 0.80$ c by means of an electric potential difference ΔV. Calculate ΔV, using the value 0.511 MeV for the rest energy (rest mass) of the electron. The MeV is the energy one electron gains when accelerated through a potential difference of 1,000,000 volts.

SOLUTION:

Procedure: Use the mass energy relation to find the energy of the electron at 0.8 c. This change in energy is the work done on the electron.

The energy, E, of the electron is given by

$$E = \frac{m_0 c^2}{\sqrt{1 - \left(\frac{v}{c}\right)^2}}$$

$$= \frac{0.511 \text{ MeV}}{\sqrt{1 - \left(\frac{0.80 \text{ c}}{c}\right)^2}} = 1.7 \times 0.511 \text{ MeV} = 0.85 \text{ MeV}.$$

The increase in kinetic energy is equal to the work done on the electron by the potential, that is,

$$\text{Work} = e\Delta V = E - m_0 c^2 = 0.36 \text{ MeV} = 0.36 \times 10^6 \times eV$$

where e is the electronic charge, so

$$\Delta V = 3.6 \times 10^5 \text{ V}.$$

The accelerating potential is 3.6×10^5 volts.

PROBLEM SG 24.8: Two electrons are traveling in the same directions with energies of 1.0 MeV and 2.0 MeV as seen from a frame S. Find the velocities of each of these electrons in S and then the velocity of the most energetic one relative to the other. Use 0.511 MeV for the electron rest mass (energy).

SOLUTION:

Procedure: Use the relativistic energy relation to find the velocities, then use the relativistic velocity addition formula.

The Einstein energy relation is

$$E = \frac{m_0 c^2}{\sqrt{1 - \left(\frac{v}{c}\right)^2}}$$

where E is the energy, m_0 is the rest mass and v is the relative velocity. Solving this for v gives:

$$\frac{v}{c} = \sqrt{1 - \left(\frac{m_0 c^2}{E}\right)^2} .$$

We let u_1 be the velocity of the slower electron and u_2 be the velocity of the faster particle. Choose the axis so that both velocities are positive. Entering the numbers gives

$$\frac{u_1}{c} = 0.86 ,$$

and

$$\frac{u_2}{c} = 0.97.$$

The relative velocity is not the difference $u_2 - u_1$; we must use the relativistic addition formula. The text formula relates the observed velocity of body 2, denoted by u, the velocity of body 1, denoted by v, and the (relative) velocity of body 2 relative to body 1 as seen by body 2, denoted by u', as

$$u = \frac{u' + v}{1 + \dfrac{u' v}{c^2}} \ .$$

Solving this for u' gives

$$u' = \frac{u - v}{1 - \dfrac{u v}{c^2}} = 0.64 \ c.$$

The relative velocity is 0.64 c.

Chapter 25
Early Atomic Structure

Important Terms

Fill in the blanks with the appropriate word or words.

1. The _____ is one event per second and is the SI unit of activity.

2. _____ is the breaking up of chemical compounds by electric currents.

3. The _____ is the normal negative charge carrier.

4. Electrons (often rapidly moving) emitted by radioactive elements are called _____ particles.

5. The smallest crystal structure is called the unit _____ .

6. The positive charges of an atom are located in the atom's _____ .

7. _____ are the short wavelength electromagnetic waves emitted by a Crooke's tube.

8. The irregular motion of minute particles suspended in a liquid or gas is called _____ motion.

9. The time it takes one–half of a radioactive substance to decay is called the _____ of the substance.

10. _____ number is how many molecules of a substance that are in one mole of the substance.

11. The _____ of a radioactive substance is the number of radioactive disintegrations per second.

12. Helium nuclei emitted in radioactive decay are called _____ particles.

13. The highly penetrating radiation from Uranium is electromagnetic. It is called _____ rays.

14. The phenomenon of emission of particles as one element changes to another is called _____ .

15. The x-rays from a Crooke's tube are also called _____ rays.

16. The relation among the wavelength of a wave incident upon a crystal, the wavelength after scattering, the crystal structure size, and the direction of scattering is known as _____ Law.

Write the definition for the following:

17. x-rays (25.5):

18. Avogadro's number (25.2):

19. electrolysis (25.2):

20. electron (25.6):

21. alpha particles (25.7):

22. beta particles (25.7):

23. gamma rays (25.7):

24. unit cell (25.1):

25. radioactivity (25.7):

26. nucleus (25.9):

27. cathode rays (25.5):

28. Bragg's Law (25.5):

29. Brownian motion (25.3):

30. activity (25.8):

31. becquerel (25.8):

32. half-life (25.9):

Answers to 1-16, Important Terms

1. becquerel; 2. electrolysis; 3. electron; 4. beta; 5. cell; 6. nucleus; 7. x–rays (or cathode rays); 8. Brownian; 9. half-life; 10. Avogadro's; 11. activity; 12. alpha; 13. gamma; 14. radioactivity; 15. cathode; 16. Bragg's

Sample Solutions

PROBLEM SG 25.1: In an electrolysis experiment a current of 2.3 amperes is maintained for 20 minutes during which zinc is deposited on the negative electrode. Zinc has an atomic weight of 65.3 and its atoms are trivalent (trivalent atoms can lose up to 3 electrons). Calculate how much zinc was deposited.

SOLUTION:

> **Procedure:** Since we know the current and the time we can calculate the total charge transported. Each atom carries an excess of 3 electronic charges, so if we divide the total charge by $3 \times e$, where e is the electronic charge, we will get the number of atoms that were transported. Since an Avogadro's number of zinc atoms has a mass of 65.3 g, a simple ratio will give the total mass.

First we calculate the total charge that flows:

$$\text{total charge} = \text{current} \times \text{time} = 2.3\ \text{A} \times 20\ \text{min} \times 60\ \text{s/min}$$

$$= 2760\ \text{C} = q,$$

where we have defined the symbol q to represent the total charge. Then we calculate the number of atoms that must flow if our calculated total charge flows:

$$\text{number of atoms} = n = \frac{\text{total charge}}{\text{charge per atom}}$$

$$= \frac{\text{total charge}}{3 \times \text{electronic charge}}$$

$$= \frac{q}{3e}.$$

There is less chance of error if we do not put numbers into the equations before the final step.

Finally the ratio defining Avogadro's number, N_A, gives,

$$\frac{\text{atomic weight of zinc}}{\text{mass deposited}} = \frac{\text{Avogradro's number}}{\text{number of atoms deposited}}$$

$$= \frac{N_A}{\dfrac{q}{3e}} = \frac{3 \times N_A \times e}{q}$$

$$= \frac{3 \times F}{q},$$

where F is the faraday.

Hence, if M is the mass deposited,

$$\frac{65.3\ \text{g/mole}}{M} = \frac{3 \times 96458\ \text{C/mole}}{2760\ \text{C}},$$

and solving for M gives

$$M = 0.62\ \text{g}.$$

A total of 0.62 grams will be deposited.

PROBLEM SG 25.2: X-rays of wavelength $\lambda = 6.73 \times 10^{-11}$ m are diffracted from a crystal. The third order scattering angle is 15° as shown in Figure SG 25.1. what is the spacing between the planes of the crystal?

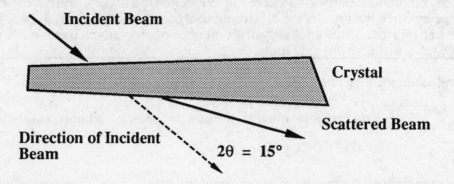

Figure SG 25.1:   X-ray scattering caused by a crystal. Note that the deviation angle is measured with respect to the incident beam and not with respect to any surfaces of the crystal.

SOLUTION:

Procedure: Equation 25.4 relates the deviation angle, the wavelength and the crystal plane spacing.

As is indicated in Figure SG 25.1, the measured deviation of the x-rays is two times the tangentially incident angle, θ. This angle must be measured from the direction of the incident beam, not from any of the crystal faces. Since we are given the scattered angle is 15°, we have $\theta = 7°.5$.

If n is the order of the scattering, λ the wavelength, d the interplanar spacing, and θ is the tangentially incident angle, then

$$n\lambda = 2\,d\,\sin\,\theta.$$

We are given n, λ, and have found θ, so we can solve for d as

$$d = \frac{n\,\lambda}{2\,\sin\,\theta} = \frac{3 \times 6.73 \times 10^{-11}\text{ m}}{2 \times \sin 7°.5} = 7.7 \times 10^{-10}\text{ m}$$

So the interplanar spacing is 7.7×10^{-10} m. Note that the factors of 2 and 3 do not affect the significant figures since they are exact numbers.

PROBLEM SG 25.3:   Electrons in a cathode ray tube are accelerated through a potential of 26.2 KV. The electron beam then passes through a pair of deflection plates 1.85 cm long separated by 0.35 cm. What potential must be applied to the plates in order to have the beam deflected 2.70 cm at a screen 25.0 cm beyond the deflection plates? The physical arrangement is shown in Figure SG 25.2. Hint: Neglect the gravitational force on the electron and the edge effects of the plates.

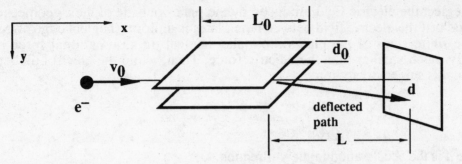

Figure SG 25.2: Electron in a cathode ray tube. The symbols are explained in the following material.

SOLUTION:

Procedure: Determine the electric force acting on the electron in terms of the applied potential. Use the laws of kinematics to calculate the resulting deflection.

This is a kinematics problem where the force is the electric field and the particle undergoing motion is an electron. We use the following notation:

Symbol	Physical quantity	value
d_0	plate separation	0.35 cm
L_0	plate length	1.85 cm
m	mass of electron	9.11×10^{-31} kg
e	charge of electron	1.60×10^{-19} C
PD_0	potential used to accelerate in x direction	26.2 KV
PD	potential used to accelerate in y direction	unknown
v_0	electron velocity entering plates	$v_0 = \sqrt{2\,PD_0/m}$
L	plate to screen distance	25.0 cm
d	linear deflection at screen	2.7 cm
F	force on electron in y direction	unknown
v	y velocity of electron after leaving plates	unknown

We make the following physical arguments:
1. While the electron is between the plates a force acts in the y-direction.
2. This force gives the electron a y-component of velocity.
3. This y-velocity produces a y-deflection at the screen.
4. We cannot use energy considerations to get the y-velocity since we cannot assume that the electron moves the entire distance from the top to the bottom plate.

The following steps are needed:
1. Find the force in the y-direction between the plates.
2. Calculate the y-acceleration between the plates.
3. Find the time spent between the plates.
4. Find the velocity in the y-direction when leaving the plates.
5. Calculate the displacement at the screen distance.

We neglect the electric field produced by the plates outside of their geometric size and assume that the electric field between the plates is uniform (neglect edge effects caused by the finite sizes of the plates). We also neglect the gravitational force since it is usually much smaller than the electric force. Then, while the electron is between the plates, it is subject to a y-force

$$eE = e\ \frac{PD}{d_0} = ma,$$

where a is the acceleration in the y-direction.

The electron enters between the plates from the left in Figure SG 25.2 and exits to the right. We know the entering x-direction velocity, v_0, from energy considerations; that is

$$PD_0 = \frac{1}{2}mv_0^2.$$

The electron is between the plates for a time t given by

$$t = \frac{L_0}{v_0} = \frac{L_0}{\sqrt{\dfrac{2e\,PD_0}{m}}}.$$

Thus, when the electron leaves the region between the plates it has a y-velocity

$$v = at = \frac{e}{m}\ \frac{PD}{d_0}\ L_0\ \sqrt{\frac{m}{2e\,PD_0}}.$$

During this time the y-displacement becomes:

$$y_0 = \frac{1}{2}\ a\ t^2 = \frac{L_0^2}{4d_0}\ \frac{PD}{PD_0}$$

We will assume this y-displacement, y_0, is small compared to the final displacement. This assumption is reasonable provided $L \gg L_0$.

If we assume that $v \ll v_0$, then it takes a time $T = L/v_0$ to travel from the plates to the screen. During this time, the y-displacement becomes

$$d = vT \quad = \frac{e}{m}\ \frac{PD}{d_0}\ L_0\ \frac{m}{2e\,PD_0}\ \frac{1}{v_0}$$

$$= \frac{PD}{PD_0}\ \frac{L_0}{d_0}\ \frac{L}{2}.$$

So

$$PD = 2\,PD_0\ \frac{d_0}{L_0}\ \frac{d}{L}.$$

The given numbers can now be entered and we find that the potential needed is 106 V. Note the advantage of not putting in the numbers until the last step. The values for e

and m are not needed. If we put the numbers in along the way we would have to multiply and divide by e and m several times. It is left as an exercise to show that

$$y_0 << d \, .$$

PROBLEM SG 22:4: A radioactive sample has a half-life of 5 min. If the initial activity is 8192 disintegrations per second, what will the activity be after one hour?

SOLUTION:

Procedure: Apply the definition of activity.

The activity is directly proportional to the number of atoms present. If the half-life is 5 min, 1 hr will be 12 half-lives. From text Table 25.2 we conclude that

$$N = \frac{N_0}{2^n}$$

where n is the number of half lives that have occurred, N_0 is the initial number of atoms and N is the remaining number of atoms. After 12 half-lives,

$$N = \frac{N_0}{4096}$$

So, if we denote the initial activity by A_0, the activity, A, after 12 half-lives will be

$$A = \frac{A_0}{4096} = \frac{8192}{4096} = 2 \, \frac{\text{decays}}{\text{s}} \, .$$

After one hour there will be 2 disintegrations/second on the average.

PROBLEM SG 25.5: A radioactive isotope decays to one-eighth its original amount in 5.0 hr. What are its half-life and decay constant? How long will it take for 10% of this isotope to decay? Suppose that, at a given time, there are 1.0×10^{15} atoms of this isotope left. What is the activity of the material?

SOLUTION:

Procedure: Count half lifes to get the half-life and then apply the definitions of decay constant and activity.

We can refer to text Table 25.3 or note that

$$\frac{1}{2^3} = \frac{1}{8} \, .$$

That is, 3 half-lifes have passed, so the half-life, $t_{\frac{1}{2}}$, is

$$t_{\frac{1}{2}} = \frac{5.0}{3} = 1.7 \text{ hr.}$$

The decay constant, λ, is given by

$$\lambda = \frac{0.693}{t_{\frac{1}{2}}} = 0.42/\text{hr}.$$

The activity A is given as

$$A = \lambda N = 0.42 \times 1.0 \times 10^{15} = 4.2 \times 10^{14} \text{ events/hr} = 1.2 \times 10^{11} \text{ Bq}.$$

PROBLEM SG 25.6: The ratio of radioactive to non-radioactive carbon in a sample of wood recovered from an archaeological dig is found to be one-fourth that of a piece of wood known to be 100 years old. Approximately how long ago was the first sample cut? Hint: Assume the natural ratio of radioactive to non-radioactive carbon for living wood has not changed in the interval.

SOLUTION:

Procedure: After each half-life there will be one-half of the radioactive carbon present as was present at the start of the half life.

If the ratio has changed to one-fourth then there have been two half-lives since the carbon in the wood was deposited in the wood. Since the half-life of radioactive carbon is about 5700 years, the old piece of wood is about 1.14×10^4 years plus the 100 years. The extra 100 years is not really needed since the age is uncertain by a larger amount.

Chapter 26
Theories and Models of the Atom

Important Terms

Fill in the blanks with the appropriate work or words.

1. The lowest energy state of a system is called the _____ state.

2. The quanta of light are called _____.

3. The _____ frequency is the minimum frequency of light which causes photoelectric emission.

4. A body whose emitted radiation depends only upon its temperature and not upon its composition is called a(n)_____.

5. The relation between wavelength of maximum emission and the temperature of an ideal body is given by the _____ law.

6. _____ law tells us that the square root of the K_α x-ray frequency is proportional to the nuclear charge.

7. The set of spectral lines of hydrogen some of which lie in the visible portion of the spectrum is called the _____ series.

8. The _____ is the minimum energy needed to remove an electron from a surface.

9. The _____ volt is the work needed to move an electron through a potential of one volt.

10 _____ is the branch of science which studies the spectra of electromagnetic radiation absorbed and emitted by substances.

11. The _____ effect is the emission of negative charge from an object as a result of light falling on the surface.

12. The ratio of the energy of a photon to the frequency of the associated light is called _____ constant.

13. The _____ formula is an empirical relation that predicts the emitted wavelength of light from hydrogen.

14. A plot representing the allowed energies of a system is called a(n) _____.

15. The _____ of a quantity is the smallest possible unit of that quantity.

16. The _____ quantum number in the Bohr theory of the atom gives the orbital size and hence the energy of an electron bound to a nucleus.

Write the definitions of the following:

17. Wien displacement law (26.4):
18. spectroscopy (26.1):

19. energy level diagram (26.7):

20. photoelectric effect (26.5):

21. photon (26.5):

22. Planck's constant (26.4):

23. principal quantum number (26.6):

24. Rydberg formula (26.2):

25. blackbody (26.3):

26. ground state (26.6):

27. quantum (26.4):

28. Moseley's law (26.8):

29. Balmer series (26.2):

30. threshold frequency (26.5):

31. work function (26.5):

32. electron volt (26.5):

Answers to 1 – 16, Important Terms

1. ground; 2. photons; 3. threshold; 4. blackbody; 5. Wien displacement; 6. Moseley's; 7. Balmer; 8. work function; 9. electron; 10. spectroscopy; 11. photoelectric; 12. Planck's; 13. Rydberg; 14. energy level diagram; 15. quantum; 16. principal.

Comments on Models

We introduce two new models of the physical world in this chapter. One of them, the Bohr model of the atom, has been replaced by models that give a more accurate description of the world. The other, the photon, describes a fundamental property of light that is used in current scientific models of the world.

The Bohr model of the atom is still used in introductory physics textbooks since it provides an easy means of visualizing of the structure of the atom. It gives very accurate results for the hydrogen atom and good results for more complicated atoms. It also provides both a historical and a physical step towards the modern wave mechanics (which is described in Chapter 27) model currently in use.

Bohr's theory suspends the rules of electrodynamics. The electron in a Bohr orbit is accelerated (circular or elliptical motion), but it does not radiate. It also suspends Newtonian laws since it is non-local information (how many wavelengths fit in an orbit) which determines the Bohr orbitals; however, the electron still remains a particle in the classical sense of being localized at a particular distance from the nucleus. Modern quantum theory modifies this model by treating the electron in an atom as a non-local object.

The photoelectric effect forces us into models which are even further from our everyday experience. Planck's photon when used to explain blackbody radiation could be considered a mathematical calculational tool; however, when Einstein explained the photoelectric effect the particle mature of light was essential. so we are faced with a model for light in which light acts as a wave (wave optics – Chapter 23) and it also acts as a particle (photoelectric effect and absorption and emission of light from atoms). This duality (wave-particle) is an essential part of the models of modern physics. a particle acts as a wave when scientists ask questions which can be answered by wave properties and it acts as a particle when we look for particle properties. We will see in the next chapter that all particles can act as waves.

The duality model of nature seems strange since we do not experience this dual behavior in everyday life. On the other hand, we should be able to accept this model provided:

A. The predictions match the experimental observations.
B. In the classical limit waves act as waves and particles as particles.

Sample Solutions

PROBLEM SG 26.1: An incandescent lamp is designed to operate at a filament temperature of 3500 K. If the lamp is operated at a higher voltage, the temperature will be higher. By what percentage will the total wavelength of maximum emission change if the lamp temperature is increased to 3700 K?

SOLUTION:

| **Procedure:** Use the Wien displacement law. |

The relevant physical law is the Wien displacement Law

$$\lambda_m T = 2.9 \times 10^{-3} \text{ m-K}, \qquad \text{(text 26.3)}$$

where λ_m is the wavelength at which the radiation distribution from a body at Kelvin temperature T peaks. For $\lambda_m(T)$ the percentage change is

$$\frac{\lambda_m(3700) - \lambda_m(3500)}{\lambda_m(3500)} \times \frac{\frac{1}{3700} - \frac{1}{3500}}{\frac{1}{3500}}$$

$$= \left(\frac{35}{37} - 1\right) \times 100 = -5\% .$$

There is about a 5% decrease in the wavelength of maximum emission.

PROBLEM SG 26.2: What is the maximum kinetic energy of the photoelectron ejected from a silver surface by ultraviolet light of wavelength 250 nm? Express this result in eV and in joules. What is the maximum wavelength that can eject photoelectrons from silver?

SOLUTION:

Procedure: Calculate the frequency for the incident light and then use the photo-electric equation to find the kinetic energy.

The photoelectron is ejected from the silver surface as shown in Figure SG 26.1. The maximum kinetic energy, KE_{max}, is given by

$$KE_{max} = hf - \phi \qquad \text{(text 26.5)}$$

where f is the frequency of the light considered as a wave, h is Planck's constant, and ϕ is the work function for silver.

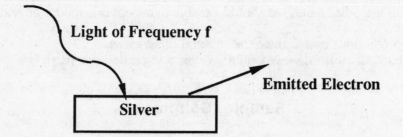

Figure SG 26.1: Photoelectron emitted from a silver surface.

From text Table 26.2 we find that $\phi_{Ag} = 4.74$ eV. If we let λ be the wavelength of the incident light, then

$$f = \frac{c}{\lambda}$$

and we can write:

$$KE_{max} = \frac{hc}{\lambda} - \phi$$

$$= \frac{6.63 \times 10^{-34} \text{ J s} \times 3.00 \times 10^8 \text{ m/s}}{250 \times 10^{-9} \text{ m}} - \phi_{Ag}.$$

$$= 7.96 \times 10^{-19} \text{ J} - 4.74 \text{ eV}.$$

Before we can subtract these two numbers they must be in the same units. The conversion is

$$1 \text{ eV} = 1.60 \times 10^{-19} \text{J},$$

so

$$KE_{max} = 0.233 \text{ eV} = 3.73 \times 10^{-20} \text{J}.$$

The work function of silver is so large that even ultraviolet light can barely liberate photoelectrons.

the maximum wavelength for photoemission will occur when

$$KE_{max} = 0.$$

This occurs when

$$\frac{h\,c}{\lambda} = \phi$$

For silver the maximum wavelength is about 262 nm.

PROBLEM SG 26.3: What is the shortest wavelength of the Brackett series ($n_1 = 4$)?

SOLUTION:

> **Procedure:** Use the Rydberg formula or the derivation of Bohr.

If we use the Bohr model, the emitted wavelength, λ, is given by

$$\frac{1}{\lambda} = \frac{m\,e^4}{8\varepsilon_0^2 h^2 c}\left(\frac{1}{(n_1)^2} - \frac{1}{(n_2)^2}\right) \quad \text{where } n_2 > n_1. \quad \text{(text 26.12)}$$

for hydrogen

$$\frac{m\,e^4}{8\varepsilon_0^2 h^2 c} = 1.097 \times 10^5 \,/\,\text{cm}.$$

Since n_1 is fixed, λ is smallest when

$$\left(\frac{1}{(n_1)^2} - \frac{1}{(n_2)^2}\right) \text{ is largest}$$

which is when $n_2 \to \infty$ ($1/n^2 \to 0$). So

$$\frac{1}{\lambda} = \frac{1.097 \times 10^5/\text{cm}}{4^2} = 6.856 \times 10^3/\text{cm}.$$

The shortest wavelength of the Brackett series is 1.459×10^{-4} cm.

PROBLEM SG 26.4: The K_α x-ray from a target has a wavelength of 0.143 nm. Which element is the target?

SOLUTION:

> **Procedure:** Use Moseley's Law to calculate the atomic number from the observed wavelength of the K_α line.

Moseley's law is given in the text as

$$f = \frac{3}{4} f_0 \, (Z-1)^2 \qquad \text{(text 26.15)}$$

where Z is the atomic number (number of protons), $f_0 = cR$ with R the Rydberg constant, $R = 109737/\text{cm}$, and f is the frequency of the K_α line.

Making the substitution

$$f = \frac{c}{\lambda} = \frac{3}{4} c R \, (Z-1)^2$$

and solving for Z,

$$Z = 1 + \frac{2}{\sqrt{3\lambda R}}$$

using our numbers (careful — 1 nm = 10^{-9} m = 10^{-7} cm);

$$Z = 1 + 29.1 \approx 30.$$

From the periodic table (text Figure 26.13) we see that the element is Zinc.

PROBLEM SG 26.5: An electron bound to a hydrogen nucleus is determined to have an orbital angular momentum of 2.13×10^{-34} J-s. What is the principal quantum number and with what energy is the electron bound to the atom?

SOLUTION:

Procedure: Use the Bohr model. The electron's angular momentum is given uniquely by the principal quantum number. Once we know the principal quantum number, we can calculate the binding energy.

The relation between the angular momentum of an electron, L, and the principal quantum number, n, for that electron is given by:

$$L = n \frac{h}{2\pi}, \qquad \text{(text 26.5)}$$

where h is Planck's constant. Since n must be an integer, it is best to substitute in the given numbers at this point and determine the proper integer.

$$n = \frac{2\pi L}{h} = \frac{2 \times 3.14159 \times 2.13 \times 10^{-34} \text{ J-s}}{6.626 \times 10^{-34} \text{ J-s}} = 2.01 .$$

Since n must be integer we take $n = 2$. This is not an unreasonable act since considering the significant figures given, the angular momentum could have been as small as 2.12×10^{-34} J-s.

Knowing n, we can calculate the binding energy from text Equation (26.11) In electron volts this equation is

$$E = -\frac{13.6\,\text{eV}}{n^2} \qquad n + 1, 2, 3, \ldots \qquad \text{(text 26.11)}$$

For $n = 2$ this relation gives

$$E = -3.4\,\text{eV}.$$

The electron is in the $n = 2$ state and is bound to the atom with an energy of -3.4 eV.

PROBLEM SG 26.6: A small meteoroid of size and mass approximately that of a golf ball (0.10 kg) is in circular orbit about the Earth. Assume that angular momentum is quantized. Calculate the principal quantum number of the orbit and the spacing between adjacent orbits. Hint: Assume the mass is uncharged so that gravity is the dominant force and that the distance of the orbit from the Earth's surface is small compared with the Earth's radius.

SOLUTION:

> **Procedure:** Equate the centripetal and gravitational forces to find how the velocity depends upon the radius. Use this relation in the quantization of angular momentum to find the value for n.

Since the meteoroid is in orbit, the gravitational force must provide the necessary force for circular motion. If we let $m = 0.10$ kg be the mass of the meteoroid, M the mass of the Earth, v the circular velocity and r the distance to the Earth's center,

$$\frac{mv^2}{r} = \frac{GMm}{r^2} = gm\left(\frac{R_E}{r}\right)^2,$$

where g is the acceleration of gravity at the Earth's surface, R_E is the radius of the Earth, and we have made use of the definition of g,

$$g = \frac{GMm}{R_E}.$$

The condition for quantizing angular momentum is

$$mvr_n = n\frac{h}{2\pi}.$$

Let us assume that the orbits must be quantized and set the r in the first equation to some r_n given by Bohr's equation.

When we combine these two relations to eliminate the velocity we get

$$r_n = \frac{n^2h^2}{4\pi^2gm^2R_E^2}.$$

Since r_n is the same order of magnitude as R_E, we can get an approximate value for n by setting $r_n = R_E$. This gives gives

$$n = \sqrt{\frac{4\pi^2 g m^2 R_E^3}{h^2}} = 5 \times 10^{43},$$

where we have used

$g = 9.8 \text{ m/s}^2, m = 0.10 \text{ kg}, R_E = 6.4 \times 10^6 \text{ m and } h = 6.64 \times 10^{-34} \text{ J-s.}$

Since we have made an approximation, the results are given to only one significant figure.

To find the interval between orbits calculate the difference in the orbital radius for $n + 1$ and for n, giving

$$\Delta r = r_{n+1} - r_n = \frac{(n + 1)^2 h^2}{4\pi^2 g m^2 R_E^2} - \frac{n^2 h^2}{4\pi^2 g m^2 R_E^2}$$

$$\approx \frac{2n h^2}{4\pi^2 g m^2 R_E^2} = \frac{2 R_E}{n},$$

where we have taken $2n + 1$ as about the same as $2n$.

So the fractional change is

$$\frac{\Delta r}{R_E} = \frac{2}{n} = 4 \times 10^{-44}.$$

This fractional difference is completely negligible and the quantum limitations can not be observed in planetary motions.

PROBLEM SG 26.7: What is the threshold wavelength for photoelectric emission from gold?

SOLUTION:

> **Procedure:** Use the photoelectric equation and the value of the work function for gold.

From Table 26.2, the work function of gold is as 5.31 eV. At threshold the maximum kinetic energy in the Einstein photoelectric equation is zero, so we have

$$0 = h f_{th} - \phi,$$

where h is Planck's constant, f_{th} is the threshold frequency, and ϕ is the work function. If λ_{th} is the threshold wavelength, this can be written as

$$h f_{th} = \frac{hc}{\lambda_{th}} = \phi,$$

or

$$\lambda_{th} = \frac{hc}{\phi},$$

so that

$$\lambda_{th} = \frac{(6.626 \times 10^{-34} \text{ Js})(3.00 \times 10^8 \text{ m/s})}{(5.31 \text{ eV})(1.60 \times 10^{-19} \text{ J/eV})} = 234 \text{ nm}.$$

The threshold wavelength for gold lies at 234 nm, which is in the ultraviolet region of the spectrum.

PROBLEM SG 26.8: Astronomers have observed a 3.0 K blackbody radiation striking the earth uniformly from all directions. This radiation is believed to be associated with the big bang which is commonly held to have occurred at the beginning of the universe. This radiation is the red shifted remnant of the initial primordial fireball. What is the wavelength of peak radiation from this fireball and how much energy per unit area and per unit time is striking the earth assuming the universe is a blackbody? From thermodynamics a slowly expanding gas without external energy being provided should obey the rule

$$TV^{1/3} = \text{constant},$$

where T is the Kelvin temperature and V is the volume occupied by a given number of "molecules" of the gas. At what wavelength did the radiation peak when the temperature of the universe was 3000 K? What was the energy density of radiation then as compared to the current epoch? Assuming the photons that makeup the blackbody radiation can be treated as a gas what volume did a typical region of the universe the size of the solar system (radius = 6.0×10^{12} m) occupy when the temperature of the universe was 3000 K?

SOLUTION:

Procedure: Use the given expansion law, the Wien displacement Law and the Stefan-Boltzmann Law (Chapter 10) to calculate the behavior.

If the temperature T is measured in Kelvin, the Wien displacement law gives us

$$\lambda_m T = 2.9 \times 10^{-3} \text{ m-K}$$

where λ_m is the wavelength of the peak in the blackbody curve.

For $T = 3.0$ K, the current wavelength peak is at 9.7×10^{-4} m, while for $T = 3000$ K the peak is at 9.7×10^{-7} m.

The Stefan-Boltzmann law is

$$R = \sigma T^4$$

where R is the emission per unit area and

$$\sigma = 5.67 \times 10^{-8} \frac{\text{J}}{\text{s m}^2 \text{ K}^4}.$$

At 3.0 K this gives

$$R_o = 4.6 \times 10^{-6} \frac{J}{\text{s-m}^2},$$

while at 3000 K, $R = 10^{12} R_o$.

To answer the last question we note that the value of a sphere of radius r is given by

$$\frac{4}{3} \pi r^3.$$

Let the subscript "o" denote the present and no subscript denote the time when T = 3000 K. We can then write

$$T_0(V_o)^{\frac{1}{3}} = T(V)^{\frac{1}{3}}$$

which simplifies to

$$T_o r_o = T r.$$

Then

$$T_o r_o = 3\text{K} \times 6.0 \times 10^{12}\, \text{m} = 3000\ \text{K} \times r$$

or

$$r = 6.0 \times 10^9\, \text{m}.$$

This is about 5 times the radius of the sun, so a volume about 5 times the solar radius has expanded to the size of the solar system as the temperature of the universe dropped from 3000 K to the present 3 K. At earlier epochs the universe was even hotter.

Chapter 27
Quantum Mechanics

Important Terms

Fill in the blanks with the appropriate word or words.

1. The _____ wavelength of a particle is inversely proportional to the mass of the particle.

2. The _____ principle relates the minimal lack of knowledge of the position of a particle and the minimal lack of knowledge of the momentum of a particle.

3. The _____ principle states that no two electrons may have their quantum numbers identical.

4. The quantum interactions between orbital angular momentum and external magnetic fields is called _____ quantization.

5. The _____ equation is a fundamental equation of quantum mechanics describing the behavior of a particle.

6. The absolute value squared of the _____ gives the probability density in quantum mechanics.

7. The ability of a particle to pass through a region of potential where it could not classically exist is called _____ .

8. The absolute square of the wave function of a particle is the _____ density of finding the particle.

9. Given the momentum of a particle there is an associated wavelength for the particle called the _____ wavelength.

10. The intrinsic angular momentum of an electron is referred to as the electron's _____ .

11. The _____ is a unit used to measure the orbital angular momentum of an electron.

12. The minimal kinetic energy that a quantum system must possess is called the _____ energy.

13. A large number of waves added so as to have an appreciable amplitude only in a small region of space is called a(n) _____ .

Write the definitions of the following:

14. de Broglie wavelength (27.3):

15. Schrödinger equation (27.4):

16. Compton wavelength (27.2):

17. spin (27.11):

18. uncertainty principle (27.5):

19. Bohr magneton (27.10):

20. Pauli exclusion principle (27.11):

21. zero point energy (27.7):

22. space quantization (27.10):

23. probability density (27.6):

24. wave function (27.4):

25. wave packet (27.5):

26. barrier penetration or tunneling (27.7)

Answers to 1–13, Important Terms

1. Compton; 2. uncertainty; 3. Pauli exclusion; 4. space, 5. Schrödinger; 6. wave function; 7. barrier penetration or tunneling; 8. probability; 9. de Broglie; 10. spin; 11. Bohr magneton; 12. zero point; 13. wave packet.

Comments on Models

Quantum mechanics is both an evolutionary and a revolutionary development from classical mechanics. It is evolutionary in the sense that it must, when the proper questions are asked, reduce to classical mechanics when Planck's constant is considered sufficiently small. On the other hand, quantum mechanics is revolutionary in as much as the basic structure of matter is different than was assumed by the classical physicists.

The wave nature of matter is probabilitistic but not in a classical sense. The wave function does not just represent a lack of accuracy; it is a basic physical indeterminacy. That is, when we give the wave function for the electron in the ground state of an atom, it is not possible to claim the electron is localized somewhere until a measurement of position is made. If this measurement is made with sufficient accuracy, the uncertainty relation forces us to say that we can't tell if the electron was bound to that atom. In general, quantum measurement must change or destroy the system, while classically the act of measurement can be done without disturbing the system under study. That is, quantum mechanics tells us that the probe particle must be considered as part of the system under study. Thus the physicist is no longer an uninvolved observer since the very act of observation effects the nature of the world.

Sample Solutions

PROBLEM SG 27.1: What is the ratio of the Compton wavelength of an electron to the Compton wavelength of a proton? If they both have the same speed, what is the ratio of the de Broglie wavelengths? (Assume all speeds are much less than the speed of light.)

SOLUTION:

> **Procedure:** Use the text formulae for the Compton and the de Broglie wavelengths.

The Compton wavelength, λ_c, of a particle of mass m is given by

$$\lambda_c = \frac{h}{mc},$$

where h is Planck's constant and c is the speed of light,. If the subscript p refers to the proton and the subscript e refers to the electron,

$$\frac{(\lambda_c)_{electron}}{(\lambda_c)_{proton}} = \frac{\frac{h}{m_e c}}{\frac{h}{m_p c}} = \frac{m_p}{m_e} = 1.84 \times 10^3,$$

while the de Broglie wavelength, λ, is given in terms of the momentum p by

$$\frac{h}{p},$$

or for small (small compared with c)velocity v by

$$\frac{h}{mv},$$

so

$$\frac{\lambda_{electron}}{\lambda_{proton}}\Big|_{same\ speed} = \frac{\frac{h}{m_e v}}{\frac{h}{m_p v}} = \frac{m_p}{m_e} = 1.84 \times 10^3.$$

The ratios are the same.

PROBLEM SG 27.2: An electron is located with an uncertainty in position of about 1.5×10^{-11} m. What is the maximum time after this measurement that the electron could be expected to be within 1.0×10^{-10} m of its original position (the approximate size of an atom), provided we assume motion only due to the uncertainty in the momentum?

SOLUTION:

> **Procedure:** Assume that the momentum of the electron is about equal to its uncertainty. Calculate how long a time it would take an electron with this momentum to travel the given distance.

The electron must obey the uncertainty relation, $\Delta x \, \Delta p_x \geq h$. Since the uncertainty in position is about 1.5×10^{-11} m, the uncertainty relation implies that:

$$\Delta p_x \geq \frac{h}{\Delta x} = \frac{6.63 \times 10^{-34} \text{ J-s}}{1.5 \times 10^{-11} \text{ m}} = 4.4 \times 10^{-23} \text{ kg m/s.}$$

An electron has a mass of 9.1×10^{-31} kg, so if we assume nonrelativistic expressions

$$\Delta v_x = \frac{\Delta p_x}{m} = \frac{\Delta p_x}{9.1 \times 10^{-31} \text{ kg}} \geq 4.8 \times 10^7 \text{ m/s.}$$

The time to travel a distance is given by

$$t = \frac{\text{distance}}{\text{velocity}} = \frac{1.0 \times 10^{-10} \text{ m}}{4.8 \times 10^7 \text{ m/s}} = 2.1 \times 10^{-18} \text{ s}$$

Since this velocity is more than 10% of the speed of light, the relativistic expression for momentum should probably be used. It is not unreasonable in such a case to use c itself as a good approximation in the uncertainty of the electron's velocity. Thus the electron could be moving at nearly the speed of light. Its position, in that case, would be given by

$$x(t) = x_0 + \Delta v \, t \approx x_0 + ct$$

so

$$t \approx \frac{x}{c} = \frac{1.0 \times 10^{-10} \text{ m}}{3.0 \times 10^8 \text{ m/s}} = 3.3 \times 10^{-19} \text{ s.}$$

With either assumption we see that the electron rapidly moves away from its original position.

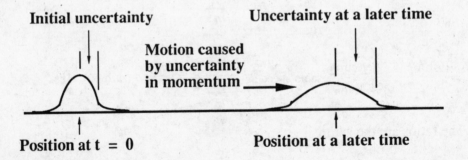

Figure SG 27.1: Minimal motion of a particle located at $x \pm \Delta x$ caused by the uncertainty relation.

Thus the electron will be on the order of 10 times its initial uncertainty away from its initial position in a time on the order of 10^{-18} s.

PROBLEM SG 27.3: Suppose an electron is confined in a very deep potential box of width $L = 0.024$ mm. What are the energy levels of the two lowest states possible? Sketch the wave functions for these two states. How does the ground state energy compare with the rest mass energy?

SOLUTION:

> **Procedure:** Apply the arguments given in Section 27.7 of the text and the equation given for the energy levels. We treat this as a one-dimensional problem.

Since the box is said to be "very deep", we assume the arguments of text section 27.7 apply and can make the sketches as shown in Figure SG 27.2.

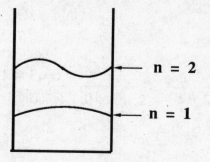

Figure SG 27.2: Wave functions of an electron in a potential box.

The energy levels are given by text Equation (27.12)

$$E_n = \frac{h^2 n^2}{8mL^2}$$

where n is the level number, h is Planck's constant, m is the mass of the particle, and L is the length of the box.

$$E_n = \left(\frac{(6.63 \times 10^{-34} \text{ J-s})^2}{8 \times 9.1 \times 10^{-31} \text{ kg} \times (2.4 \times 10^{-5} \text{ m})^2}\right) n^2$$

$$= 1.0 \times 10^{-28} \, n^2 \text{ J}.$$

The rest mass energy of an electron is

$$E_0 = m c^2 = 9.1 \times 10^{-31} \text{ kg} \times (3.0 \times 10^8 \text{ m/s})^2 = 8.2 \times 10^{-15} \text{ J}.$$

So the ratio of the ground state energy to the rest mass is about 10^{-14}.

PROBLEM SG 27.4: In the preceding problem, what must the value of L be such that the ground state energy is equal to the rest mass energy of the electron?

SOLUTION:

> **Procedure:** Find the value of L in the preceding expression for E_n that makes $E_n = E_0$ where E_0 is the rest mass energy.

We have two relationships:

$$E_0 = m_e c^2$$

and

$$E_n = E_{binding} = \frac{h^2 n^2}{8 m_e L^2}.$$

where E_0 is the rest mass energy of an electron with mass m_e, c is the speed of light, E_n is the binding energy for quantum number n, h is Planck's constant, L is the box dimension, and n is an integer from 1 to any number.

Equating these for $n = 1$ gives,

$$L = \frac{1}{2\sqrt{2}} \frac{h}{m_e c} = \frac{6.63 \times 10^{-34} \text{ J-s}}{2.83 \times 9.1 \times 10^{-31} \text{ kg} \times 3.0 \times 10^8 \text{ m/s}}$$

$$= 8.6 \times 10^{-13} \text{ m}.$$

This size, 8.6×10^{-13} m, is about 20 times the typical size of a nucleus. This again shows us that electrons, as individual particles, cannot be localized in an atomic nucleus.

PROBLEM SG 27.5: How many hydrogenic electronic states are there corresponding to the principal quantum number $n = 4$?

SOLUTION:

> **Procedure:** Count the possible orbital quantum numbers, l, for $n = 4$ and count how many magnetic quantum numbers, m_l, correspond to each l. Multiply each number by two to allow for two spin states.

For each value of the principal quantum number n, the orbital quantum number l may assume integral values between 0 and $n - 1$ (n values). For each value of l, the magnetic quantum number, m_l, may assume any integral value between $-l$ and $+l$ ($2l + 1$ values. Finally for each value of m_l, the electron may have its spin up or down, $m_s = \pm 1/2$. The table below catalogs the possible states for n = 4.

l value	m_l values	m_s values	number of states
3	$-3, -2, -1, 0, +1, +2, +3$	$\pm 1/2$	14
2	$-2, -1, 0, +1, +2$	$\pm 1/2$	10
1	$-1, 0, +1$	$\pm 1/2$	6
0	0	$\pm 1/2$	2
		Total number of states	32

There are 32 electronic states for $n = 4$.

PROBLEM SG 27.6: A magnetic storm on the Sun associated with a cluster of sunspots is observed to produce a Zeeman splitting of 1.2×10^{-4} eV as shown in Figure SG 27.3. What is the magnetic field that produced this splitting?

$$\Delta E = 1.2 \times 10^{-4} \text{ eV}$$

Figure SG 27.3: Zeeman splitting of hydrogen spectral line produced by the solar magnetic field associated with a sunspot cluster.

SOLUTION:

> **Procedure:** Calculated the energy difference in the $\pm 1/2$ spins produced by a magnetic field. Solve for the field.

The energy of each state is given by

$$E = \mu_s B = 2\mu_B m_s B,$$

so the energy difference is

$$\Delta E = E_+ - E_- = 2\mu_B \left\{ \frac{1}{2} - \left(-\frac{1}{2} \right) \right\} B = 2\mu_B B$$

$$= 2 \times 9.27 \times 10^{-24} \frac{\text{J}}{\text{T}} \times \frac{1}{1.6 \times 10^{-19}} \frac{\text{eV}}{\text{J}} \times B$$

$$= 1.16 \times 10^{-4} \frac{\text{eV}}{\text{T}} \times B.$$

Hence,

$$B = \frac{1.2 \times 10^{-4} \text{ eV}}{1.16 \times 10^{-4} \frac{\text{eV}}{\text{T}}} = 1.0 \text{ T}.$$

The associated magnetic field was 1.0 T.

PROBLEM SG 27.7: Electrons are individually accelerated through a potential difference of 1000 volts. Each passes through a slit of width d, and strikes a detecting screen located a distance 0.50 m away. Over a period of time a diffraction pattern is produced. What should the slit size be in order that the first minimum in the electron diffraction pattern on the screen be 0.10 mm from the center of the pattern?

SOLUTION:

> **Procedure:** Calculate the momentum from energy conservation and use this momentum to calculate a de Broglie wavelength. Use this wavelength to calculate the first minimum in the diffraction pattern.

If V is the potential difference, e the electronic charge, m the electron's mass, and p the momentum of the electron before it passes through the slit, we have from energy conservation:

$$eV = \frac{p^2}{2m},$$

provided we assume that the velocity is non-relativistic. You should verify this assumption with the numbers given.

So

$$p = \sqrt{2meV}.$$

The de Broglie wavelength λ is given by

$$\lambda = \frac{h}{p},$$

so

$$\lambda = \frac{h}{\sqrt{2meV}}.$$

The first minimum in the diffraction pattern for a wavelength λ, a deflection angle θ, and slit size d was given in Chapter 23 as

$$\sin \theta = \frac{\lambda}{d} = \frac{y}{D},$$

where y is the deflection of the electron and D is the distance to the screen as shown in Figure SG 27.4. The last equality follows from the assumption that $y \ll D$.

The width is thus

$$d = \frac{\lambda D}{y} = \frac{hD}{y \sqrt{2meV}}$$

$$= 1.9 \times 10^2 \text{ nm.}$$

This is a slit width about the size of the wavelength of visible light and only several hundred times larger than the spacing between planes of molecules in a solid.

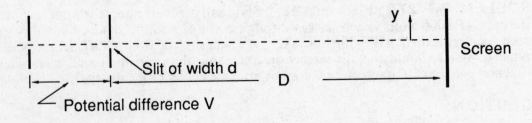

Figure SG 27.4:   Electron diffraction.

Chapter 28
The Nucleus

Important Terms

Fill in the blanks with the appropriate word or words.

1. The neutral particle in the nucleus is the _____.

2. _____ are nuclei with the same charge but different numbers of neutrons.

3. A(n)_____ has the same mass as an electron, but has opposite charge.

4. The dose _____ takes into account not only the energy deposited per unit mass by radiation but also the biological effect of the type of radiation.

5. Nuclear fissions induced by neutrons released by previous nuclear fissions are called a chain _____ if more and more fissions are induced as time passes.

6. The _____ is the measure of the ability of a particular type of radiation to cause biological damage.

7. The _____ is the energy deposited per unit mass by absorbed radiation.

8. A(n) _____ occurs when an orbital electron is captured by its nucleus.

9. The minimum energy that a particle needs in order to react with a nucleus is called the _____ energy.

10. When two nuclei combine to form another, we have the process known as _____.

11. The _____ of a nuclear decay is the difference between the initial and final mass energies.

12. The positive charges in the nucleus are _____.

13. Protons and neutrons are collectively referred to as _____.

14. The difference between the sums of the mass energies of the nucleons in a nucleus and the mass energy of that nucleus is termed the _____ of that nucleus.

15. _____ plus and _____ minus are other names for positrons and electrons respectively.

16. The _____ is a particle with nearly zero rest mass and no charge. It has the same spin as the electron.

17. The process in which one nucleus breaks up into two or more nuclei is termed _____.

Write the definitions of the following:

18. proton (28.1):

19. chain reaction (28.10):

20. neutron (28.1):

21. nucleon (28.2):

22. isotope (28.2):

23. dose equivalent (28.8):

24. positron (28.6):

25. beta plus (β^+) and beta minus (β^-) (28.6):

26. quality factor (28.8):

27. electron capture (28.6):

28. neutrino (28.6):

29. threshold energy (28.9):

30. fission (28.10):

31. fusion (28.11):

32. dose (28.8):

33. Q-value (28.4):

34. nuclear binding energy (28.3):

Answers to 1 – 17, Important Terms

1. neutron; 2. isotopes; 3. positron; 4. equivalent; 5. reaction; 6. quality factor 7. dose; 8. electron capture; 9. threshold; 10. fusion; 11. Q-value; 12. protons; 13. nucleons; 14. binding energy; 15. beta (β^+ and β^-); 16. neutrino; 17. fission.

Comments on Models

The major portion of this chapter deals with the models used to describe nuclear decay. Three different but related models are discussed. These give a description of α, β, and γ-decay.

The model for α-decay involves the classical and quantum ideas of a potential well. That is, we assume the α-particle is able to tunnel out of the nucleus. The remaining nucleons in the nucleus are used to define this potential well that retains the α-particle. The β-decay model invokes energy conservation to force us to accept the idea of the neutrino. The neutrino is given the properties needed to conserve energy during β-decay reactions. Until recently one of the properties of the neutrino was that it had no mass. Recent experiments hint that the neutrino may have a small mass. Remember that the terms mass, spin and charge are only properties that we assign to our models. The γ-decay model assumes that the nucleus has energy levels analogous to those in the electronic structure of atoms. In this model the entire nucleus is assumed to have these energy levels. some nuclear physicists do spectrometry of these nuclear energy levels and attempt to deduce the nuclear structure in the same fashion that atomic physicists do spectrometry on atomic levels to deduce the electronic structure of the atom. All three of these models are used to make predictions as to how nuclei will behave.

The α and γ- decay models are similar since they both can be considered as potential well problems, but they also have their fundamental differences. In the α-decay case all the nucleons together define the potential well. At this point in our studies we are where the atomic spectroscopists were before the Bohr theory was devised. Chapter 31 will help to resolve some of these ideas.

Sample Solutions

PROBLEM SG 28.1: Natural potassium occurs as a mixture of three isotopes:

$$^{20}_{10}\text{K at } 93.08\%, \quad ^{21}_{10}\text{K at } 0.01\% \text{ and } \quad ^{22}_{10}\text{K at } 6.91\%.$$

Calculate the atomic mass of naturally occurring potassium assuming that the atomic mass of any isotope is given by the number of nucleons in the nucleus. (this assumption is good to better than 1-2%).

SOLUTION:

> **Procedure:** Naturally occurring potassium should consist of the given mixture of isotopes. Calculate the weighted average of the masses of the three isotopes. The weighting used is the percentage of occurrence.

The general procedure to use if there are N isotopes of masses m_i and each having a percentage occurrence of P_i, i = 1 to N; is to to write

$$m_{observed} = \frac{\sum\limits_{i=1}^{N} m_i P_i}{\sum\limits_{i=1}^{N} P_i} = \frac{\sum\limits_{i=1}^{N} m_i P_i}{100}.$$

For potassium we write

$$m_{observed} = \frac{20 \times 93.08 + 21 \times 0.01 + 22 \times 6.91}{100} = 20.1 \text{ u.}$$

the atomic mass of naturally occurring potassium is nearly the same as that of the most abundant isotope.

PROBLEM SG 28.2: One of the decays of an isotope of polonium is

$$^{208}_{84}Po \rightarrow \, ^{204}_{82}Pb + \, ^{4}_{2}\alpha \, ,$$

where the lead (plumbium) isotope has an atomic mass of 204.0363 u. The emitted alpha particle has a kinetic energy of 5.108 MeV. What is the atomic mass of the polonium isotope, assuming no other particles are emitted? What is the Q-value of this decay?

SOLUTION:

Procedure: Add the rest energy of the α particle to its kinetic energy to get the total energy in the α particle. This energy plus the mass energy in the Pb nucleus must be the total initial energy in the Po nucleus. Use conservation of momentum and energy to determine how this energy is distributed.

We can used text Table 28.2 to help us answer this question. The initial mass of the Po must be the sum of the final Pb mass plus the final kinetic energy of the Pb nucleus plus the mass of the a-particle plus the kinetic energy of the alpha particle. That is:

mass of Po = mass of Pb + mass equivalent of Pb kinetic energy
+ mass of α + mass equivalent of α particle kinetic energy.

Text Table 28.2 gives the α particle mass as 3728.431 MeV, so the total energy associated with the α particle is

$$(3728.431 + 5.108) \text{ MeV} = 3733.539 \text{ MeV}.$$

In order to compare masses we must be able to convert mass in u to mass in Mev and conversely. This conversion factor can be calculated by comparing the quoted value of the mass of the proton to the quoted value in MeV. This ratio is

$$\frac{\text{mass of p in u}}{\text{mass of p in MeV}} = \frac{1.007276}{938.279} \frac{u}{\text{MeV}}$$

$$= 1.0735 \times 10^{-3} \frac{u}{\text{MeV}} \cdot$$

The emitted alpha particle has an equivalent mass of

$$3733.539 \text{ MeV} \times 1.0735 \times 10^{-3} \frac{u}{\text{MeV}} = 4.0081 \text{ u}.$$

This gives two of the four terms we need (mass of α + mass equivalent of α particle kinetic energy) To calculate the kinetic energy term we will need to consider the recoil of the Pb nucleus and conserve momentum and energy. The decay occurs as shown in Figure SG 28.1 where v_α is the velocity if the alpha particle, v_{Pb} is the velocity of the Pb nucleus, m_{Pb} is the mass of the lead nucleus, and m_α is the mass of the alpha particle

Conservation of momentum gives

$$v_{Pb} \, m_{Pb} + v_\alpha \, m_\alpha = 0,$$

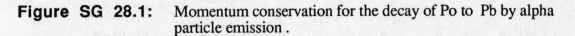

Figure SG 28.1:   Momentum conservation for the decay of Po to Pb by alpha particle emission .

Since the kinetic energy of the α particle is small compared with the rest energies, all velocities will be small compared with the speed of light, so we can use the non-relativistic kinetic energy expression:

$$\text{(Kinetic Energy)}_{Pb} = \frac{1}{2}m_{Pb}v_{Pb}^2 = \frac{1}{2}m_{Pb}\left(\frac{m_\alpha}{m_{Pb}}v_\alpha\right)^2$$

where the last equality results from substituting in the expression for conservation of momentum. Thus

$$\text{(Kinetic Energy)}_{Pb} = \frac{1}{2}m_\alpha v_\alpha^2 \frac{m_\alpha}{m_{Pb}} = \frac{m_\alpha}{m_{Pb}} KE_\alpha$$

$$= 0.01961 \times 5.108 \text{ MeV}$$

$$= 0.100 \text{ MeV}.$$

The mass equivalent of this energy is

$$0.100 \text{ MeV} \times 1.0735 \times 10^{-3} \frac{u}{\text{MeV}} = 10^{-4} \text{ u}.$$

Therefore,

$$\text{mass of Po} = 204.0363 \text{ u} + 0.0001 \text{ u} + 4.0081 \text{ u} = 208.0445 \text{ u}.$$

So $^{208}_{84}$Po has a mass of 208.0445 u.

The Q-value is the sum of the two kinetic energy terms so

$$Q = (5.108 + 0.100) \text{ MeV} = 5.208 \text{ MeV}.$$

PROBLEM SG 28.3: A small radioactive source is monitored with a detector that subtends a solid angle of 0.080 steradians. If the average count rate in the detector is 7396 counts per minute, what is the activity of the source in Bq?

SOLUTION:

> **Procedure:** Assume the radiation is emitted spherically. Calculate the fraction of a sphere that the detector intercepts. Multiply the counts/s by the inverse of this fraction to get the activity.

The geometry of this problem is shown in Figure SG 28.2.

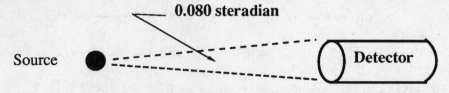

0.080 steradian

Figure SG 28.2: Detector with a 0.080 steradian intercept angle.

We must assume the radiation is emitted isotopically in all directions. That is, the radiation is emitted uniformly over the entire sphere which is 4π steradians. This means

$$\frac{\text{total counts/min}}{4\pi \text{ steradians}} = \frac{\text{observed counts/min}}{0.080 \text{ steradians}}$$

So

$$\text{Total counts/min} = \frac{4\pi}{0.080} \times 7396 = 1161761 = 1.16 \times 10^6 \text{ counts per minute.}$$

The activity in Bq is the counts per second so

$$\text{activity} = \frac{1.16 \times 10^5 \text{counts per minute}}{60 \text{ seconds per minute}} = 1.9 \times 10^4 \text{ Bq.}$$

The activity is about 20,000 Bq.

PROBLEM SG 28.4: A proton with a kinetic energy of 5.00 MeV elasticity scatters head-on off a deuteron which was initially at rest. What is the kinetic energy of the deuteron in MeV after the collision? Use the mass for the deuteron atom given in text Table 28.2.

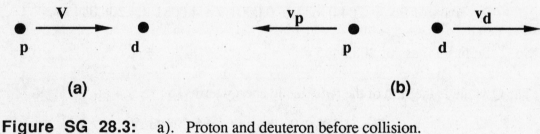

Figure SG 28.3: a). Proton and deuteron before collision.
b). Particles after collision.

SOLUTION:

Procedure: Since the kinetic energy of the proton is small compared with its rest energy, this is a non-relativistic problem. Conserve energy and momentum.

This is a collision problem identical to those done in the mechanics portion of this book. The main difference is that energies are expressed in MeV instead of J. The term "head-on" implies the collision can be treated as a one dimensional problem. Let the subscript "p" denote the proton and a subscript "d" denote the deuteron. Let V be the initial velocity of the proton and v and m denote velocities and masses; then, conserving momentum and energy as shown in Figure SG 28.3, we have:

$$m_p V = m_p v_p + m_d v_d \qquad \text{(momentum conservation)}$$

$$\frac{1}{2} m_p v^2 = \frac{1}{2} m_p v_p^2 + \frac{1}{2} m_d v_d^2 \qquad \text{(energy conservation).}$$

Solve the momentum conservation equation for v_p and eliminate v_p from the energy equation. This gives:

$$2 V = \left(1 + \frac{m_d}{m_p}\right) v_d$$

or

$$v_d = \frac{2V}{1 + \left(\frac{m_d}{m_p}\right)}.$$

We can check our calculations to this point by noting that if $m_d = m_p$ then $v_d = V$ and $v_p = 0$ as expected.

Finally

$$KE_d = \frac{1}{2} m_d v_d^2 = \frac{1}{2} m_d \times \left(\frac{2V}{1 + \left(\frac{m_d}{m_p}\right)}\right)^2$$

$$= 4 \frac{m_d m_p}{(m_p + m_d)^2} KE_p$$

where KE_p is the initial kinetic energy of the incoming proton. Thus,

$$KE_d = 0.889 \times KE_p = 4.44 \text{ MeV.}$$

The scattered deuteron has a kinetic energy of 4.44 MeV.

PROBLEM SG 25.5: Determine the kinetic energy available to the alpha decay of

$$^{235}_{92}U \rightarrow ^{231}_{92}Th + ^{4}_{2}\alpha$$

if the observed masses are

$$m\left(^{235}_{92}U\right) = 235.0494 \text{ u,} \quad \text{and} \quad m\left(^{231}_{92}Th\right) = 231.03631 \text{ u.}$$

SOLUTION:

Procedure: The decay must conserve energy.

We assume the decay products are non-relativistic and write:

mass of U = mass of Th + kinetic energy of Th
+ mass of alpha + kinetic energy of alpha.

Thus

kinetic energy of decay products

= kinetic energy of Th + kinetic energy of alpha

= mass of U − mass of Th − mass of alpha

$$= 235.0494 \text{ u} - 231.03631 \text{ u} - 3728.431 \text{ Mev} \times 1.0735 \times 10^{-3} \frac{u}{MeV}$$

$$= 0.0106 \text{ u} = 0.0106 \text{ u} \times \frac{1}{1.0735 \times 10^{-3} \frac{u}{MeV}} = 9.87 \text{ MeV}.$$

This kinetic energy will be divided between the Th and the α. The division is determined by conservation of momentum. See Problem SG 28.4.

PROBLEM SG 28.6: Determine the identity of the nucleus X in the reaction:

$$^{13}_{6}C + n \rightarrow X + \gamma.$$

SOLUTION:

Procedure: Reactions must conserve charge and mass numbers.
The reaction may become more clear if we write the mass and charge numbers for the neutron and the γ ray.

$$^{13}_{6}C + ^{1}_{0}n \rightarrow X + ^{0}_{0}\gamma.$$

Then we see that the nucleus X has a mass number of $13 + 1 - 0 = 14$ and a charge number of $6 - 0 - 0 = 6$. Thus the decay product is carbon 14.

Chapter 29
Lasers, Holography, and Color

Important Terms

Fill in the blanks with the appropriate word or words. The various types of holograms are included for identification.

1. The emission of light by excited atoms triggered by the presence of other light of the proper frequency is termed _____ emission.

2. When more ions are in an excited state than are in the lower state we have a(n) _____.

3. The measure of an optical instrument's ability to produce distinct images of objects which appear close together is called the _____ of the instrument.

4. A(n) _____ medium contains atoms or molecules which can be made to emit coherent light.

5. A(n) optical _____ contains mirrors that reflect light back and forth many times.

6. A(n) _____ hologram is viewed by observing its image with nearly monochromatic light which is passed through the hologram.

7. The inverse of the number of lines per unit length that can be resolved by a holographic film is termed the _____.

8. The coherence of light extending in space is called _____ coherence.

9. A(n) _____ hologram can be viewed with a broad wavelength source provided the source is nearly a point source.

10. A photographic recording of an interference pattern is called a(n) _____.

11. The making and study of the photographic recording of interference patterns is called _____.

12. The light emitted by the random de-excitation of atoms is known as _____ emission.

13. The _____ is an optical example of stimulated emission.

14. A nearly monochromatic light bean is said to have a high _____ coherence.

15. _____ holograms are made by stamping the interference pattern into plastic.

16. _____ is the term for the process of exciting the laser medium.

17. _____ holograms are produced by having the reference and object beams strike the opposite sides of the film.

18. The _____ hologram is a transmission hologram which may be viewed with white light. The color of the holographic images varies as the angle of viewing is changed.

19. It is possible to create and maintain inverted populations in an active medium because there exists a(n) _____ state with a longer lifetime than might be expected.

20. The _____ primary colors of light are red, green, and blue.

21. The _____ primary colors are those which combined with one of red, green, or blue produce white light.

Write the definitions of the following:

22. holography (29.5):

23. laser (29.1):

24. active medium (29.2):

25. inverted population (29.2):

26. spatial coherence (29.4):

27. additive primary colors (29.7):

28. spontaneous emission (29.1):

30. pumping (29.2):

31. metastable state (29.2):

32. hologram (29.5):

33. subtractive primary colors (29.7):

34. stimulated emission (29.1):

35. optical resonator (29.2):

36. temporal coherence (29.4):

Answers to 1–22, Important Terms

1. stimulated; 2. inverted population; 3. resolving power; 4. active; 5. resonator;
6. transmission; 7. spatial frequency; 8. spatial; 9. white light; 10. hologram;
11. holography; 12. spontaneous; 13. laser; 14. temporal; 15. embossed; 16. pumping;
17. reflection; 18. rainbow; 19. metastable; 20. additive; 21. subtractive

Sample Solutions

PROBLEM SG 29.1: The interference pattern produced by light from a laser falling on two slits has a fringe visibility of 0.60. What is the ratio of the maximum to the minimum intensity? Assuming that one of the slits is illuminates two times as strongly as the other, what is the coherence?

SOLUTION:

> **Procedure:** Use the definitions of fringe visibility and coherence to calculate the requested quantities.

If I_{max} is the maximum intensity and I_{min} is the minimum intensity, then the visibility V is defined in the text as

$$V = \frac{I_{max} - I_{min}}{I_{max} + I_{min}}.$$

This can be rearranged to give

$$\frac{I_{max}}{I_{min}} = \frac{1 + V}{1 - V}.$$

$$= \frac{1 + 0.60}{1 - 0.60} = 4.0$$

The coherence is defined as

$$\gamma = \frac{I_1 + I_2}{2\sqrt{I_1 I_2}} V,$$

where I_1 and I_2 are the illumination at slits 1 and 2 respectively and V is the visibility. This gives

$$\gamma = \frac{3I_1}{2\sqrt{2I_1^2}} V = \frac{3}{2\sqrt{2}} 4.0 = 4.2.$$

The intensity ratio is 0.40 and the coherence is 4.2.

PROBLEM SG 29.2: Two beams of parallel laser light with a wavelength, λ, of 632.8 nm are brought together on a photographic plate to form a sinusoidal holographic grating. One beam strikes the plate at normal incidence. The other beam makes an angle of $32^\circ.0$ with the normal. (See Figure SG 29.1.) The processed hologram is illuminated with light of wavelength 488.0 nm. If this light strikes the hologram at normal incidence, how many beams emerge and at what angles?

SOLUTION:

> **Procedure:** Calculate the spacing of the grating. This can be combined with the formula for a diffraction pattern from Chapter 23 to get the maximum order permitted by the grating.

If one laser beam (the reference) strikes the film normally and the other strikes at an angle θ_0 with respect to the normal as shown in Figure SG 29.1, then the interference pattern formed has an effective grating spacing of d given by

$$d = \frac{\lambda_1}{\sin \theta_0} \; ,$$

where λ_1 is the wavelength of the laser light.

If this diffraction grating is now illuminated normally with laser light of wavelength λ_2, diffraction maxima will occur at (See Chapter 23) angles θ_n given by

$$\sin \theta_n = \frac{n \lambda_2}{d} = n \frac{\lambda_2}{\lambda_1} \sin \theta_0,$$

for $n = 1, 2, 3, \ldots, n_{max}$ where the largest n possible, n_{max} is that for which $n_{max} + 1$ would make the right hand side greater than 1. Entering our numbers gives

$$\sin \theta_n = n \frac{488 \text{ nm}}{632.8 \text{ nm}} \sin 32° = 0.409 \, n.$$

Since n = 3 makes the right hand side greater than 1, there are two diffraction orders whose angles can be found by taking the inverse sine; that is,

$$\theta_n = \sin^{-1} 0.409 \, n \, .$$

The diffraction maxima occur at $24°.1$ and at $54°.8$.

Figure SG 29.1: Two laser beams combining to form a sinusoidal holographic grating.

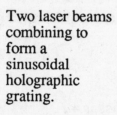
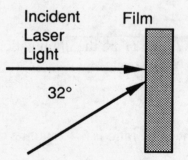

PROBLEM SG 29.3: The temperature of the star Sirus is about 10,000 K. At what wavelength is the most light emitted. What would be the color of this star?

SOLUTION:

Procedure: Use the Wien displacement law to find the wavelength of maximum emission and compare this with the visible color wavelengths.

The Wien displacement law of Chapter 26 gives us the wavelength of maximum emission, λ_{max}, in terms of the Kelvin temperature, T, as

$$\lambda_{max} = \frac{0.00290 \text{ K-m}}{T} = 2.90 \times 10^{-7} \text{ m} = 290 \text{ nm.}$$

This is shorter than visible light (400 nm $< \lambda_{visible} <$ 700 nm), so we would expect the star to look blue. Since the human eye's response drops off towards the blue, Sirius appears blue-white.

PROBLEM SG 29.4: A photon has a measured energy of 1.77 eV. What color sensation would this photon produce?

SOLUTION:

Procedure: Compute the wavelength corresponding to this energy to decide the color sensation.

To convert energy in eV to energy in joules, multiply by the electron charge, e, (magnitude — not the negative number). Let E_0 denote the energy in eV and E the energy in joules, then

$$E = E_0 e.$$

The frequency, f, of this light is given by

$$f = \frac{E}{h},$$

and the wavelength, λ, by

$$\lambda = \frac{c}{f},$$

where c is the speed of light.

Combining gives

$$\lambda = \frac{ch}{eE_0} = 7.01 \times 10^{-7} \text{ m} = 701 \text{ nm.}$$

This is at the long wavelength end of the visible spectrum, so the sensation would be deep red.

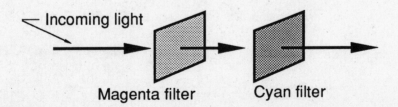

Incoming light

Magenta filter   Cyan filter

FIGURE SG 29.2: White light passing through a magenta filter followed by a cyan filter

PROBLEM SG 29.5: White light passes through a magenta filter, which is, in turn, followed by a cyan filter as shown in Figure SG 29.2. What color emerges?

SOLUTION:

Procedure: Determine what colors are passed by the first filter and then which of these will be passed by the second filter.

Both of these filters are subtractive primary color filters. The magenta filter passes the colors complementary to green and the cyan filter passes the colors complementary to red.

The magenta filter removes the green color. Thus white light minus green enters the cyan filter. The cyan removes the red color. The two filters together have removed both the red and the green. Only the blue is left.

The light will appear blue.

Chapter 30
Condensed Matter

Important Terms

Fill in the blanks with the appropriate word or words.

1. A(n) _____ crystal has a regular periodic arrangement in only one or two dimensions.

2. The energies of electrons in a crystal are only permitted to be in regions called _____.

3. The constant of proportionally between the applied electric field and the resulting current density is called the electrical _____.

4. A pure material with a conductivity value between that of a good metal and a good insulator is called a(n) _____ semiconductor.

5. A solid forming a three-dimensional array or structure is said to be _____.

6. The energy of the most energetic electron in a substance at 0 K is called the _____ energy.

7. Semiconductors whose properties depend upon the addition of a small number of other atoms are called _____ semiconductors.

8. A solid made up of many individual crystalline subunits is called a(n) _____ solid.

9. A(n) _____ semiconductor is made by adding impurities which donate negative carriers to the conduction band.

10. The _____ are those values of energies which electrons in a solid can not have.

11. A solid without any long range order is called a(n) _____ solid.

12. The _____ is a measure of the drift velocity given electrons in a solid per unit applied electric field.

13. An insulator with a small energy gap between the last filled band and the empty conduction band becomes a(n) _____.

14. If a magnetic field and an electric field at right angles to the electric field are both applied to a conductor, the _____ voltage is generated at right angles to both the magnetic field and the electric field.

15. A(n) _____ semiconductor is made by adding impurities which trap electrons out of the conduction band.

Write the definitions of the following:

16. amorphous solid (30.1):

17. electrical conductivity (30.2):

18. forbidden zones (30.4):

19. Hall effect (30.6):

20. p-type semiconductor (30.7):

21. crystalline solid (30.1):

22. liquid crystal (30.1):

23. mobility (30.3):

24. semiconductor (30.4):

25. impurity (extrinsic) semiconductors (30.7):

26. polycrystalline solid (30.1):

27. Fermi energy (30.2):

28. allowed bands (30.4):

29. pure (intrinsic) semiconductors (30.5):

30. n-type semiconductor (30.7):

Answers to 1–15, Important Terms

1. liquid; 2. allowed bands; 3. conductivity; 4. pure (intrinsic); 5. crystalline; 6. Fermi;
7. impurity (extrinsic); 8. polycrystalline; 9. n-type; 10. forbidden zones; 11. amorphous;
12. mobility; 13. semiconductor; 14 Hall; 15. p-type

Sample Solutions

PROBLEM SG 30.1: Palladium (Pd) has an atomic mass of 106.7 u, a mass density of 1.22×10^4 kg/m^3. Assume that solid palladium has 2 free electrons per atom. Calculate the Fermi energy of Palladium.

SOLUTION:

> **Procedure:** Use the provided numbers to calculate the number of atoms per unit volume and then the free electron density.

We will let ρ represent the mass density of Palladium, m_{Pd}, the mass density of Palladium, n_e the number of free electrons per atom. The the free electron density, n, is given by

$$n = n_e \times \frac{\rho \times N_A}{m_{Pd}},$$

where N_A is Avogradro's number. Entering numbers, and remembering to convert atomic mass units to kg/mol, we find that

$$n = 2 \times \frac{1.22 \times 10^4 \text{ kg/m}^3 \times 6.02 \times 10^{23} \text{ atoms/mol}}{1.067 \times 10^{-1} \text{ kg/mol}}$$

$$= 1.38 \times 10^{29} / \text{m}^3.$$

Using this value of n, we can find the Fermi energy as

$$E_F = \frac{h^2}{2m}\left(\frac{3n}{8\pi}\right)^{\frac{2}{3}}$$

$$= 1.56 \times 10^{-18} \text{ J} = 9.76 \text{ eV}$$

The Fermi energy of Palladium is 9.76 eV.

PROBLEM SG 30.2: The electrical resistivity of Palladium is 10.2×10^{-8} Ω-m. What is the electronic mobility and the collision time for the conduction electrons?

SOLUTION:

Procedure: Use the free electron density from Problem SG 30.1 and the electrical conductivity to calculate the collision time for the carriers. From this find the mobility.

The resistivity is defined as the reciprocal of the conductivity. The conductivity σ, is given in terms of the free electron density n, the electronic charge e, the electron mobility μ, and the electron mass m as

$$\sigma = ne\mu,$$

which can be solved for

$$\mu = \frac{\sigma}{ne}$$

$$= \frac{\dfrac{1}{1.02 \times 10^{-7} \, \Omega\text{-m}}}{1.38 \times 10^{29} / \text{m}^3 \times 1.60 \times 10^{-19} \text{ C}}$$

$$= 4.44 \times 10^{-4} \frac{\text{m}^2}{\text{V-s}}.$$

The collision time is given by

$$\tau = \frac{\mu m}{e}$$

$$= \frac{4.44 \times 10^{-4} \frac{m^2}{V\text{-}s} \times 9.1 \times 10^{-31} \text{ kg}}{1.60 \times 10^{-19} \text{ C}}$$

$$= 2.53 \times 10^{-15} \text{ s}$$

The electron mobility in Palladium is $4.44 \times 10^{-4} \frac{m^2}{V\text{-}s}$ and an electron will suffer a collision about every 2.53×10^{-15} seconds.

PROBLEM SG 30.3: A 100 foot extension cord is made of copper wire. It is carrying sufficient current that at 2.0 V drop occurs along each direction of the cord. Given the electrical resistivity of copper as 1.7×10^{-8} Ω-m and that copper has 8.45×10^{28} free electrons per cubic meter, how long will it take on the average for an electron to travel one way on the wire? Hint: Assume that this is a dc current and a dc potential difference.

SOLUTION:

Procedure: The resistivity allows us to calculate the collision time. The time of travel for the electron is determined by the electric field and the average velocity.

The average drift velocity v is given in terms of the electric field E and the collision time τ as

$$v = \frac{eE}{m} \tau$$

Since the electric field is

$$E = \frac{PD}{L}$$

where PD is the potential difference and L is the length of the wire and since

$$\tau = \frac{\sigma m}{ne^2},$$

where σ is the conductivity, m is the electron mass, e is the electron charge, and n is the free electron density; we can write

$$v = \frac{\sigma PD}{neL}.$$

The time of travel t will be given by

$$t = \frac{L}{v} = \frac{neL^2}{\sigma PD}$$

$$= \frac{8.45 \times 10^{28} \times 1.6 \times 10^{-19}\ \text{C} \times (30.5\ \text{m})^2}{\dfrac{1}{1.7 \times 10^{-8}\ \Omega\text{-m}} \times 2.0\ \text{V}}$$

$$= 1.1 \times 10^5\ \text{s}.$$

This result may seem surprising. The current shows up almost immediately in the conductor; however, it will take an individual electron about 30 hours to travel the length of the extension cord. Clearly we have a cooperative effect where the electrons we use are probably not the ones we put into the cord from the power supply.

The preceding calculation ignores the heating effect on the conductivity which would result in a smaller wire having an even smaller drift velocity. With an ac power supply the drift velocity is not changed by much in magnitude (provided $\tau \ll 1/\text{frequency}$) and the reversal of the voltage means that the electrons from the power lines never get to what ever we have attached to the cord. You are buying power not electrons from the power company.

PROBLEM SG 30.4: A ribbon of p-type semiconductor has a z-dimension of 1.5 mm and a y-dimension of 4.0 mm. The ribbon is suspended in a magnetic field of strength 1.5 T in the z-direction. The material has a free carrier density of 3.2×10^{22} carriers/m^3. What is the Hall voltage if 200 mA is passed in the x-direction through the sample?

SOLUTION:

> **Procedure:** Use the relation defining the Hall field given in the text.

Let l_z represent the z-dimension of the ribbon, l_y the y-dimension, B_z the magnetic field, i the current in the x-direction and n the density of carriers as shown in Figure SG 30.1.

Figure SG 30.1: Semiconductor ribbon in an external magnetic field.

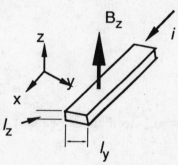

The Hall field, E_y is given by

$$E_y = \frac{j_x B_z}{en}$$

$$= \frac{iB_z}{l_z l_y en} \,,$$

where j_x is the current density and e is the charge on the carriers. Since this is a p-type semiconductor, $e > 0$, so E_y points in the $+y$-direction.

The Hall voltage is then

$$V_y = \frac{iB_z}{l_z l_y en} \times l_y = \frac{iB_z}{l_z en}$$

$$= \frac{2.0 \times 10^{-1} \text{ A} \times 1.5 \text{ T}}{1.5 \times 10^{-3}\text{m} \times 1.6 \times 10^{-19} \text{ C} \times 3.2 \times 10^{22} \text{ m}^{-3}}$$

$$= 0.039 \text{ V}.$$

The Hall voltage will be 39 mV in the $+y$ direction.

PROBLEM SG 30.5: What size area of solar cells are needed to produce 50 kW if the incident solar flux is 1.0 kW-m^{-2} and the solar cells are 20% efficient.

SOLUTION:

> **Procedure:** The power produced is proportional to the area and the incident energy and the efficiency of the cell.

If we let P be the desired power, A the area of the solar cell array, E the efficiency, and I_0 the incident power per unit area, then

$$P = AEI_0.$$

The required area is

$$A = \frac{P}{EI_0}$$

$$= \frac{50 \text{ kW}}{0.20 \times 1.0 \text{ kW-m}^{-2}}$$

$$= 250 \text{ m}^2.$$

You would need an array about 16 m on a side.

Chapter 31
Sub-Nuclear Forces

Important Terms

Fill in the blanks with the appropriate words or words.

1. Particles just like the electron except for their charge are called _____.

2. The _____ has a mass of about 170 times that of the electron and has the same spin.

3. Particles which interact through the strong nuclear force are called _____.

4. _____ theories attempt to explain the basic forces of nature in terms of a single force.

5. The fundamental building blocks of atoms are called _____ particles.

6. The _____ is a particle with no spin and a mass larger than the muon. It was first predicted by Yukawa.

7. The _____ are intermediate mass particles which may be produced singly in nuclear reactions.

8. Protons and neutrons are believed to be composed of three _____.

9. _____ are about a thousand times as massive as electrons. When they decay they produce and other light particles but never baryons.

10. The elementary particles consist of hadrons which experience the strong nuclear force and photons and _____ which do not. The electron is an example of the latter.

11. The _____ theory has the origin of the universe 10-20 x 10^9 years ago.

12. _____ are about a thousand times as massive as an electron and always include a baryon in their decay products.

13. _____ are the heavy particles which feel the strong nuclear force. They can be produced in pairs only.

14. A particle which has the same mass of another particle but a different charge is termed a(n) _____. Such pairs will annihilate then they meet.

Write the definitions of the following:

15. muon (31.2):

16. kaon (31.3):

17. hadron (31.4):

18. lepton (31.4):

19. unified theories (31.6):

20. big bang (31.7):

21. positron (31.1):

22. antiparticle (31.1):

23. pion (31.2):

24. hyperon (31.3):

25. meson (31.4):

26. baryon (31.4):

27. quark (31.5):

28. elementary (or fundamental) particle (31.0):

Answers to 1-17, Important Terms

1. positrons; 2. muon; 3. hadrons; 4. unified; 5. fundamental (or elementary); 6. pion;
7. mesons; 8. quarks; 9. kaons; 10. leptons; 11. big bang; 12. hyperons; 13. baryons;
14. antiparticle

Comments on Model

The quark model described in the text applies to the hadrons. It provides us a picture of
how the hadrons are interrelated. The quark model appears to bring us back to the point
particles or classical physics. This is not a completely correct view. The interaction of these
particles is basically nonclassical.

Sample Solutions

Handy conversion factors:

1 MeV corresponds to a mass of 1.7827×10^{-29} kg

$$1 \text{ MeV} = 1.60 \times 10^{-13} \text{ J}$$

PROBLEM SG 31.1: A proton and an antiproton collide directly. Each is moving
towards the other at a speed of 0.010 times that of light. Assuming that these velocities are
non-relativistic, what is the total energy released in MeV when these two particles meet and
annihilate? In what form (number of photons) would you expect the radiation to appear?

SOLUTION:

> **Procedure:** Calculate the total energy (rest + kinetic) before the collision. The initial energy must all be converted to light. The produced photons must satisfy the symmetry of the problem.

(a)   **(b)**

Figure SG 31.1:   a) Colliding proton and antiproton b) The reaction products.

Since we are told to assume non-relativistic motion for the particles we should be able to write:

$$E_p = m_0c^2 + KE_p$$

and

$$E_{\bar{p}} = m_0c^2 + KE_{\bar{p}},$$

where p denotes the proton and $\bar{p}$ the anti-proton. We assume these two have the same rest mass energy and that we can use the value of 938.279 Mev given in text Table 28.2 for the rest mass term.

Since the particles have the same rest mass, conservation of momentum implies that the two kinetic energies are equal; so we can write

$$KE_{total} = m_0v^2 = m_0c^2 \frac{v^2}{c^2} = 938.279 \text{ MeV} \times \left(\frac{0.010 \text{ c}}{c}\right)^2$$

$$= 938.279 \text{ MeV} \times 10^{-4} = 0.094 \text{ MeV}.$$

The total energy is then

$$E_{total} = 2 m_0c^2 + KE_{total} = 1.876652 \times 10^3 \text{ MeV}.$$

The reaction products must be one or more photons. One is not possible since the initial system of particles had no net momentum before collision. The minimal possible solution must consist of two photons of equal energy (and hence frequency) traveling in opposite directions. Although we have not covered all the material needed for the complete solution, it should be pointed out that the two photon result is only possible provided the spins of the two particles are opposed. That is, to end up with two photons, it is necessary to assume that the two particles form a singlet s state in analogy to the s state of atomic spectra.

PROBLEM SG 31.2: A π^o (neutral pion) at rest in the laboratory decays into two photons. What are the frequencies of the photons? Assume the speed of light is exactly 3.000×10^8 m/s.

SOLUTION:

> **Procedure:** Assume energy conservation and symmetry. The two photons have the same frequency and their total energy must be the rest energy of the pion.

The reaction can be written as

$$\pi^o \rightarrow \gamma_1 + \gamma_2.$$

Since there is no momentum before the decay, we expect the photons will have equal energy and be directed oppositely to each other as shown in Figure SG 31.2.

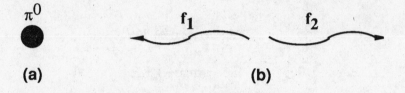

(a) **(b)**

Figure SG 31.2: a) Neutral pion at rest. b) Final decay photons.

The energy of a photon is given by hf where h is Planck's constant and f is the frequency of the light. Thus, conserving energy and letting m_o be the pion rest mass:

$$m_o c^2 = 2hf \quad \text{or} \quad f = \frac{m_o c^2}{2h}.$$

We can get the mass of pion from text Table 28.1; however. it is in MeV. The conversion is 1 MeV corresponds to 1.7827×10^{-29} kg. So the mass of the pion is

$$m_o = 135.0 \times 1.7827 \times 10^{-29} \text{ kg} = 2.407 \times 10^{-27} \text{ kg}.$$

Then the frequency is

$$f = \frac{2.407 \times 10^{-27} \text{kg} \times (3.000 \times 10^8 \text{ m/s})^2}{2 \times 6.63 \times 10^{-34} \text{J-s}}$$

$$= 1.64 \times 10^{23} \text{ Hz}.$$

This frequency corresponds to very hard gamma rays.

PROBLEM SG 31.3: What is the de Broglie wavelength of a 20 GeV electron?

SOLUTION:

> **Procedure:** Use the relation defining the de Broglie wavelength.

The de Broglie wavelength was given is text Chapter 27, Equation (27.2) as

$$\lambda = \frac{h}{p}$$

where h is Planck's constant and p is the momentum of the particle.

We are given the kinetic energy of the electron, 20 GeV = 20×10^3 MeV. Since this is much larger than the energy associated with the rest mass (rest mass energy = 0.511 MeV) we can neglect the rest mass. The relativistic relation between mass and energy [text Equation (24.12)], then becomes

$$p \approx \frac{E}{c}$$

so

$$\lambda = \frac{hc}{E} .$$

We must put the energy in joules [1 MeV = 1.60×10^{-13} J], so

$$\lambda = \frac{6.63 \times 10^{-34} \text{J-s} \times 3.00 \times 10^8 \text{m/s}}{2.0 \times 10^4 \times 1.60 \times 10^{-13} \text{J}}$$

$$= 6.22 \times 10^{-17} \text{ m} .$$

The wavelength is a very short 6.22×10^{-17} m .

PROBLEM SG 31.4: What is the charge of a particle whose quark composition is d$\bar{\text{s}}$? Is this a baryon or a meson?

SOLUTION:

> **Procedure:** Add the individual quark charges.

The d quark has a charge of $-1/3$ while the $\bar{\text{s}}$ has a charge of $+1/3$ so this particle has a charge of zero. Since it is composed of only two quarks, the particle must be a meson.

PROBLEM SG 31.5: Can the reaction $p + p \rightarrow p + n + \pi^+$ occur? Explain why or why not.

SOLUTION:

> **Procedure:** Reactions must satisfy charge conservation, energy conservation, spin conservation, and baryon number conservation.

To determine if a decay can occur we must check to see if it satisfies conservation laws. In addition to the three conservation laws from previous chapters (energy, charge, spin), this chapter adds a fourth conservation law (baryon number).

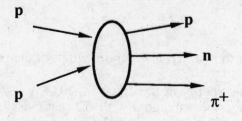

Figure SG 31.3:   The reaction $p + p \rightarrow p + n + \pi^+$.

Charge conservation: The right hand side has a total charge of $+2e$ as does the left hand side. Charge conservation is satisfied.

Energy conservation: Energy conservation is not relevant in this case since we have two incoming particles which we are allowed to give any relative velocity. This allows the energy on the right hand side to be as large as needed to satisfy energy conservation. Energy conservation is very useful if there is only a single particle at the start.

Spin conservation: The proton and the neutron are spin 1/2 particles while the pion is a spin 0 particle. The spins on the right hand side can add to 1 or 0 as can the spins on the left hand side so spin conservation can be satisfied.

Baryon number conservation: The proton and neutron each have a baryon number of $+1$ and the π^+ has a baryon number of 0. The right hand side has a baryon number of $+2$ as does the left hand side. Baryon number conservation is satisfied.

The reaction can occur.